The Spacious Days

The Spacious Days

Michael F. Twist

FARMING PRESS

ISBN 0 85236 241 2

A catalogue record for this book is available
from the British Library

Published by Farming Press Books
Wharfedale Road, Ipswich IP1 4LG, United Kingdom

Distributed in North America
by Diamond Farm Enterprises,
Box 537, Alexandria Bay, NY 13607, USA

Cover photos
Front: *Philip Wadman and part of the Burnham Hampshire
Down flock, Lynch Hill Lane, 1938.
The end of Lamas Wood is in the background.*
Back: *The author on his fifth birthday in 1924 with
the Hereford bull Quartz.*

Cover design by Mark Beesley
Typesetting by Galleon Photosetting, Ipswich
Printed and bound in Great Britain by
Redwood Books, Trowbridge, Wiltshire

Contents

	Preface	*vii*
1	Of Bulls and Boundaries	1
2	Tuppence a Pint	15
3	Early Days with the Gun	26
4	Pin Money	40
5	A Spotted Pig and Garnet	52
6	Pheasants and Gundogs with a Difference	62
7	Hay and Harvest	74
8	Holidays	89
9	Poachers and Poaching	104
10	Grass Snakes and Squirrels	121
11	Threshing	133
12	The Flock	145
13	Opportunities	159
14	The End of an Era	171

**A section of photographs
appears between pages 120 and 121**

Dedicated to my daughter Diana,
for whom this book was written,
and
to my brother, Ralph,
who was so much a part of those spacious years

Preface

The Spacious Days is an account of my early life, through the twenties and thirties, on an agricultural estate at Burnham in Buckinghamshire, now largely buried beneath a sea of houses. It tells of my love of roaming the countryside, early involvement in farming, and unchildlike passion for cattle, particularly bulls and other livestock.

The book recounts stories of pet foxes, 'pet' stoats, not so friendly, a grey squirrel that became a household pet and a badger that thought it was a lap dog!

At the time about which the book is written a three bedroomed cottage could be built for well under £300, beer cost 2d per pint and 6d (2½p) was considered adequate pocket money for the young. Rats were a serious menace to crops, but they had a monetary value, as did other pests, for they carried a 'bounty'. I supplemented my pocket money by catching rats at 2d per tail, whilst moles and queen wasps were worth 6d each.

The book takes one back to the time when a farm labourer often worked twelve to fourteen hours a day and did so through pride in his work, as well as a desire for extra money. There are tales of poachers and general villainy. The latter includes an account of an ingenious farm worker who developed a unique method of stealing rabbits from the harvest field.

Days are recalled holidaying in North Devon in August and early September, where miles of beaches without a soul were to be seen and where large catches of prawns and, indeed, lobsters were easily caught when the tide was out. The scarcity of traffic was such that, in one village through which I was regularly driven by my father in a bull-nosed Morris, an aged inhabitant could have his large armchair placed in the middle of the village street so as to be able to enjoy the morning sunshine!

It was a time when a small estate such as at Burnham (about 1,500 acres) provided employment for some forty workers, with jobs for a further twenty in the 'big' house and gardens. Today such a holding would, in total, employ no more than five or six people. The book tells of haymaking and harvesting, when horses were the main source of power, of country pursuits and a retriever being unable to do its job through being drunk on Drambuie! These and numerous other anecdotes, about real people and real places, provide an insight into days long gone – days before man had really made his mark on unbalancing nature and on the destruction of wildlife and the environment as a whole.

MICHAEL F. TWIST
Woolpit, Suffolk

1

Of Bulls and Boundaries

FROM a very early age I never had any doubts as to what I wanted to do in the future. It was to be a land agent-cum-farm manager like my father, who held that position on the Burnham Grove Estate in South Bucks for about thirty years. We moved to Burnham on 3 July 1921, my second birthday.

Burnham at that time was a country village, whose history could be traced back to the Domesday Book. Small, quiet and self-contained, it offered much in the way of human and natural wealth long since devoured by the insatiable appetite of progress. Many of them were little things, like fresh crisp bread, still hot from the oven, delivered to the house before eight o'clock in the morning; the High Street resounding to the whistling of errand boys as they sped about their business; the pungent smell of burning hoof, wafting from the smithy, as a shoe was fitted, or the huffing and puffing of Cleare's Foden Steam Wagon as it made its way down the street to deliver coal and anthracite, its solid tyres bumping over the potholes. Butcher, grocer, fishmonger, chemist, draper, builder, doctor, architect, even a foundry and a brickyard – you name it and it was there. A complete unit, built slowly and laboriously over many centuries. But time marches on and, to a great extent, that proud and singular autonomy has gone. Houses and factories replace fields and orchards; the forty-minute drive to London, largely through rural countryside and pleasant villages like Colnbrook and Osterley, has been replaced by a mad dash up the M4. Farms and market gardens, where once I accompanied my father as a guest to shoot partridges, are buried beneath the might of London Airport.

So much change and so much devastation since that warm summer's day seventy years ago when I arrived at my new home, situated some three hundred yards from the bottom of the High Street and at the top of Hog Fair Lane opposite the old village Pound. In fact the house was known as Pound Cottage, having once been two cottages that were knocked into one and to which various additions had been made. Looking south across the fields the nearest building we could see was Windsor Castle. Alas the house has long since disappeared, as has the garden in which my father took such pride, to be replaced by a host of houses. The adjoining field has been developed as a school and playing fields, the only view a sea of roof tops.

Pound Cottage was about a quarter of a mile from Britwell Farm, the hub of the estate's activities. There were, eventually, four other farms – Lynch Hill, Biddles, The Leas and the Home Farm. The latter adjoined Burnham Grove, the home of the estate's owner, Edward Clifton-Brown, a partner in the merchant banking firm of Brown, Shipley & Co. In addition there were the Sheep Sheds, a set of buildings larger than many farmyards, reserved entirely for the Hampshire Down flock which, by the end of the 1930s, was the leading one in the country. The estate covered an area that engulfed most of the land between Burnham village to the west, the Farnham Royal – Slough road to the east, the Great Western Railway line to the south and Burnham Beeches to the north. The latter was and, indeed, still is a beech forest of great antiquity, which over the years had been much favoured by various monarchs, when at Windsor, for hunting stags.

My enthusiasm and desire to follow in father's footsteps led me into trouble on more than one occasion. All farm livestock fascinated me, particularly cattle. Nearly every day, when I was between three-and-a-half and four years old, I was taken to Britwell to see the cows if it was fine. One day I was missing. Nanny, having frantically searched the house and garden, alerted mother. They scoured the spinney in front of the house, the field adjoining and even went some distance along the footpath that ran past the house, eventually to join up with the private road leading to Lynch Hill and Biddles. Not a sign. Mother, I was to learn years later, was frantic when father returned from somewhere on the estate. He did not delay and headed straight for Britwell. I was quickly discovered in one of the cow byres, sitting happily on the floor under a large Shorthorn cow and making a very passable job of milking her! Father extracted me, protesting vigorously. My protests changed to cries of anguish as I received a sound spanking. I knew very well that I was not supposed to go outside the garden – in fact I couldn't unless someone opened the gate – but an errand boy making a delivery to the house had done just that, leaving it open long enough for me to make my escape.

It was not many months later that I was again in trouble. Several of the farmhands, when they could be spared from other jobs, were helping to remake the garden. Two of these had done service in the Boer War and one, old Bill Herbert, was reputed to have fought in the Zulu War. They were a source of never-ending entertainment to my elder brother, Ralph, and me. Their stories were as fruity as their language and old Bill, in particular, was vituperative on the subject of army N.C.O.s to a degree that I have seldom heard equalled. Both Ralph and I increased our vocabularies considerably.

In those days it was customary to have afternoon tea at

3

about four o'clock – not a mug of tea and a biscuit, but a sit-down meal with scones, thinly sliced bread and butter, comb honey, jam, a variety of dainty little sandwiches and an assortment of cakes – all, of course, home made. On the day that I was to prove that I had been keeping my ears open, the village doctor's wife, Mrs Croly, a somewhat autocratic lady, came to tea. Ralph and I were scrubbed, dressed up in our best shorts and shirts, ties neatly knotted and delivered pristine and irritable by nanny to the drawing room. Here we received a 'briefing' from mother – be polite and only speak when spoken to. It was a hot summer and I can remember to this day what Mrs Croly was wearing – a pale pink dress, white shoes and stockings and a vast straw hat sitting squarely on her head with more artificial fruit on it than I have ever seen adorning anyone's headgear other than the film star Carmen Miranda's! Our visitor totally ignored my brother and me, in spite of my mother saying, 'You know the boys.' It was not until I was busily stuffing my mouth with bread and butter, liberally plastered with strawberry jam, that she deigned to address me.

'Well, my little man, and what are you going to do when you grow up?'

I ignored her and was quickly reprimanded by mother. I scowled across at our guest, then replied: 'I'm not your little man and it's none of your bloody business but, if you must know, I'm having Daddy's job when he's dead.'

It seemed as though the world had exploded around me. I was literally dragged from the room by mother, still clutching my bread and jam, rushed upstairs to nanny and, before you could say Jack Robinson, was being thoroughly 'paddled' with the back of the hairbrush. Meanwhile mother had raced back to Mrs C., who was having the vapours – she never came to tea again! I was put to bed in disgrace and remained there with no supper.

It was very hot and the windows of my bedroom were wide open. Later that evening I could hear mother and father talking as they walked on the lawn. Mother was giggling.

'You know it was terribly naughty of Mick, but I do sympathise, she is very condescending and, I find, most difficult to get on with.' Father laughed.

'Oh, she's alright. I hope he learned a lesson from having his backside warmed, namely that you cannot always say exactly what you think.'

'You should have seen her face. Where do you think he learned such language?'

'Oh, from the men working in the garden. They're pretty fluent – and if he's been near old Bill you were lucky it wasn't worse.' Father was right, I did learn from that incident to hold my tongue and say 'nowt', something that has stood me in good stead on many an occasion since then.

Shortly after the debacle of Mrs Croly I again escaped to Britwell. On this occasion I excelled myself. The maintenance staff were painting the buildings and had left a pair of steps leaning against the wall, right by one of the bull-boxes. The bull housed in this particular loose-box was a Jersey. He was so bad tempered that the herdsman at the Home Farm, where Mrs Clifton-Brown had a small Jersey herd, refused to look after him and he was moved to Britwell. The temptation was too great. I climbed up the steps. The bull, Mickens Rob Roy, was lying down. I leaned over the half-door, trying to get him to come to me, but he was apparently not interested. In my endeavours to persuade him I leaned a little too far and the next thing I knew I was inside, sprawled on the floor.

Fred Simpson, the head herdsman, just saw me disappearing over the door. In later years he told me it was one of the most terrible moments of his life. To his credit

5

he did not panic, but ran across the yard as quietly as he could. There was no noise coming from the loose-box, so he cautiously peered in. Rob Roy was still lying down, fortunately not far from the door, and I was stroking his nose! Hardly daring to breathe Fred eased back the bolts. Then he half-opened the door, grabbed me by the arm and almost hurled me into the yard, slamming the door and just getting one bolt home as an irate bull hit it with an almighty crash. I had no idea how lucky I had been. Fred was far more concerned than I was as he returned me to Pound Cottage. Strangely I did not get into trouble. I think both my parents were too relieved that I was not just a splodge on the wall to worry about retribution for my sins. After some consultation, to which I was not privy, it was agreed with Fred that, whenever feasible, I should go down to the farm during the afternoon and spend time with him.

So started a period of sheer bliss. In the morning my brother and I were taught the three Rs by a governess from 9.00 a.m. until noon. (We lunched at that hour to fit in with the working of the estate, for that was the time that the staff had their midday break. The dairymen were at work by 5.00 a.m., the head carter at each farm by 6.00 a.m., so as to have the horses watered and fed when the rest of the men came in at 7.00.) Providing it was fine, as soon as it was time I headed for Britwell. To begin with someone escorted me, but it was not long before I was allowed to go on my own, with strict instructions to be back in time for tea. These days the thought of letting a four-and-a-half to five-year-old child out on the road, on his own, is unthinkable, but in the early twenties there was virtually no traffic and in the country the fear of being molested was non-existent.

When I first started my regular visits to the farm, apart from a few milk cows, the main herd was pedigree

Herefords, shortly to be sold. However, before this happened there was time for me to develop a great rapport with the stock bull – Quartz. He was very gentle. I frequently had a ride on him, but what I enjoyed even more was to hold the ends of the lead ropes when Fred took him for his daily exercise. One day I discovered Fred had let go and I was actually leading Quartz! My joy and importance knew no bounds, and it was not long before I was taking him for his daily walk around the orchard adjoining the farm, as Fred followed with one of the other bulls. It is a complete myth that all bulls are savage. They are not, and particularly in some of the beef breeds many, like Quartz, are as docile as kittens. On my fifth birthday, prior to my party, Mr Ferris the photographer from the village met father and me at Britwell and I was photographed with my beloved Quartz. I still have that photo and wept bitter tears a few weeks later when the herd was dispersed and Quartz went to a new home.

However, I quickly got over my grief in all the excitement surrounding the formation of the new Red Poll herd. Edward Clifton-Brown's roots were in East Anglia and he was anxious to have a herd of what was his 'native' breed. Every day that I went to the farm I was learning and it was not long before I could mix the rations for the cows. In those days there were no balanced feeds available from merchants, one mixed one's own. The foods that one could purchase, other than cereals, were limited. Apart from home-grown beans the chief source of protein was imported linseed cake. This came in slabs approximately 3 feet long, 18 inches wide and about 1 inch thick. Before it could be used it had to be put through a crusher. Cotton seed cake came from Egypt in a similar form.

Tom Bunce, better known as 'Peg Leg', for he had lost a leg when still a child and had a wooden one, was in charge of the granary built onto the end of the old timbered barn at Britwell. Every Friday, first thing after lunch, this was opened. The big Ruston Hornsby engine started up and cake would be crushed, oats rolled and barley and beans ground. In the evening a number of the staff came in and bagged and weighed off the rations for each farm and sheep unit. These three to four hours' overtime were much sought after, particularly in winter time, to a degree that a number of the men worked on a rota. At that time the basic agricultural wage for the county was around 32 shillings a week (£1.60 in present-day money), but men were paid according to their ability and some earned as much as £2 10s a week, or even £3 in the case of the foreman. Senior members of the staff got a free cottage, milk and in some cases a hundredweight of coal a week.

With an arable flock of sheep, that is one that is folded over roots and other crops, there was a large acreage of swedes and turnips to be flat-hoed and singled, as well as mangolds for the cattle. Much of the singling, that is spacing out the plants, was done piecework and many of the men on a fine summer's morning, would be in the fields by 4.00 a.m. to put in three hours' hoeing before starting their day's work. I was about twelve before I was allowed to earn any money this way, although I had by then spent many hours both flat-hoeing and thinning roots. I well remember my first morning. I was in the field, Hunger Hill, by 5.00 a.m. The rate was good – 1s 2d per hundred yards. (It varied according to the density of the crop, anything from 10d to 1s 3d.) By the time the four men, who were in the field before I arrived, went off to start their day's work, my back was stiff and my hands blistered, but I stuck at it until a quarter to eight. Although it would be checked by the foreman, Tom Rose, I stepped out what I had done – 450 yards. That meant I'd earned 5s 3d (about 26.5 p in present terms), a fortune when one received a shilling a week pocket money, half of which had to be saved. I raced home on my bike for breakfast, my hoe tied to the crossbar with a bit of old binder-twine, just like a real farm labourer. I was elated. Making money was simple, you just had to get up and go out and work for it! However, much was to happen before that memorable time.

I worked hard in the mornings under the watchful eye of our governess, although from the age of six I was on my own. Ralph had gone to stay with a great friend of my mother's and attended a school as a day-boy; both of us were to go as boarders when I was eight (a terrifying thought which I managed normally to dispel from my mind). I became quite knowledgeable, for one so young, on the subject of sex. When I was at Britwell and a cow was bulling and had to be served, I would sit up in one of the many four-wheeled wagons and watch. One day I asked Fred Simpson what was happening.

'Oh,' came the reply, ' the cow and the bull are having a wedding.' This satisfied me for a while – and then I asked why.

'So they can have a baby. All animals have to do it, including people.' Fred was slightly hesitant in his reply but, at the time, the full significance of his biology lesson did not sink in. Later I gave the matter great thought.

Some months afterwards, Edward Clifton-Brown's daughter, Rhona, was married. It was the event of the year as far as Burnham village was concerned and caused much local interest. Mother and father were officially invited. I was taken to the church by Kate, who had replaced nanny as maid-cum-child minder, to see all the splendour of the occasion. The church was packed. At some solemn moment, when one could have heard a pin drop, a piping little voice rose from the back of the church.

'Kate, this isn't anything like the weddings the cows and bulls have at Britwell.'

I never knew how far up the church my observations were heard. Certainly, in our immediate vicinity, there were gasps of horror, as faces turned in our direction. A blushing Kate hurried me from the church, home and straight to bed, where I remained in disgrace without tea or supper. For once I was at a loss to know what I had done wrong – I'd merely stated what, for me, was a fact!

Worse was to come. I was banned from going to Britwell for a month, which didn't worry me too much as I had become very friendly with Bob Hedges, the second gamekeeper, who was secretly teaching me to shoot with an airgun. However, events outran the ban. Two weeks later, around midday, I started getting excruciating pain in the right side of my abdomen. Dr Weaver Adams, our doctor from Slough, came within the hour and diagnosed appendicitis. An immediate operation was necessary. Mother, a fully trained nurse, wished me to remain at home. Can you imagine such a proposition even being considered today? However, Dr Adams contacted Sir Joseph Skeffington, the leading surgeon in the area, and he had no objection. So early that evening I became the largest joint ever carved on our dining room table. Dr Adams gave the anaesthetic, a mixture of chloroform and ether dripped onto a pad of cotton wool in a small mask covering my mouth and nose. Sir Joseph, aided by a sister from Windsor hospital and mother, completed the operation with the minimum of fuss, leaving a large scar as evidence – it was no case of keyhole surgery! For a week I was not even allowed to sit up in bed – so different from today – but went downstairs for tea on my seventh birthday.

My recovery was rapid and it was not long before I was again heading for Britwell, or joining Bob Hedges as he walked around his beat. The big excitement was the forthcoming arrival of a Red Poll bull, Bredfield Nathan.

He was no ordinary bull, for he was one of the top sires in the breed, and had also attained great success in the show-ring. But what made him particularly interesting to a small boy was the fact that he had nearly killed the herdsman who looked after him and had left the poor man crippled for life. Father heard that Nathan, because of his misdeeds, was to be slaughtered. After a quick consultation with Fred, an offer was made for Nathan: twice his meat value plus the cost of transport, providing his current owner would arrange delivery to Britwell. This was accepted and a new loose-box was hastily constructed. It had an entrance either end and a sliding door to divide it, worked by pulleys from the outside, so that the occupant could be shut in whichever section was desired. There was even a small hatch over the feeding trough, so at no time would it be necessary for anyone to enter the box.

The new bull-box was complete and the day of Nathan's arrival fixed. In those days the transportation by road of around one ton of bull from near Woodbridge in Suffolk to South Bucks was, in itself, something of an achievement. At last he arrived. My excitement was intense. I took up my position in a wagon, from where I would be able to see straight into the lorry when the ramp was let down. As this happened the noise, half-bellow, half-roar, that came from the inmate was spine chilling. I had never heard such fury! Father and Fred were masterminding the unloading, aided by about ten men. There seemed to be a confusion of ropes, then suddenly, with a tremendous bellow, Nathan was out. The plan was that there should be five men on each rope, one well out on either side, so that he could not get at anyone. However, Nathan's sudden descent had caught the 'leaders' off their guard. There was only one on each rope, albeit for just a matter of seconds, but that was long enough! With an almighty bellow he charged blindly, crashing into the rear

11

wheel of the wagon in which I sat, rocking it and half turning it as he did so. He raised his head and we were virtually eyeball to eyeball. Never before, or since, have I seen such venom and hate in an animal's eyes. Hesitantly I reached out and touched his head. I don't remember feeling fear, just a strange magnetism towards this magnificent creature. The 'leaders' had by then got themselves organised and Nathan was manoeuvred towards his new home.

Later, when he was safely ensconced and all was quiet, I asked Fred if I could get Nathan some food. The answer was in the affirmative and I rushed off for a large scoop of meal and several handfuls of linseed cake. I slid back the feeding hatch and looked in. Nathan was in the further section, pawing the ground and sending straw flying into the air. I called him. With a snort he turned and came lumbering up to the hatch, grumbling as he did so. I held out some linseed cake and, after a moment's uncertainty, a big rough tongue scooped it off my hand. I fed him the rest and then emptied the feed I had brought into his trough, talking to him while he was eating and for some while after he had finished. Slowly the hate went out of his eyes. Suddenly his great rough tongue came out and licked my arm as I leant against the hatch. It was like being rubbed with wet sandpaper! I was delighted. I turned to Fred, who had been standing nearby, quietly watching.

'He's lovely. I know we'll be friends. I don't think he's as bad as he's made out to be.'

'You're probably right, but don't forget he's not Quartz. We mustn't take any risks and it'll take a long, long time to get him round.'

I went home to lunch elated. I had a new friend and Fred had said 'we'. That meant that I was involved in Nathan's future.

The following year passed all too rapidly. I worked hard at my lessons and gloried in my afternoons at the farm. I had little interest in toys, except my model farm, and read avidly anything I could get my hands on connected with cattle. Life was good. Nathan was becoming more and more appeased and Fred could now enter his loose-box without first attracting him to the hatch with food and tying him up. I regularly groomed him when secured, standing on a bucket so that I could reach the top of his shoulders. Once, when Nathan lowered his head, I stepped off this and sat astride his neck, rather like a mahout. All he did was raise his head, lifting me high in the air, and continue chewing the cud. Fred, who was mucking out the further section, was not amused. He removed me from my lofty perch and, rightly, severely reprimanded me.

Suddenly, it seemed without warning, the dreaded day arrived and I found myself in the car heading towards boarding school. Details, thankfully, are now blurred, but what recollections I have are of a Dickensian establishment. The porridge at breakfast was a congealed slab that one could spin intact on its cold plate. We had to wash in midwinter, stripped to the waist, in front of wide open windows, with snow and rain often blowing in. But above all else I remember the two principals, sisters, who must surely have inspired the writing of *Arsenic and Old Lace*.

Nothing is more boring than other people's illnesses but, alas, I must briefly record that in my four terms at boarding school I developed pneumonia three times and never completed a term, returning home each time by ambulance. I had not really fully recovered from this – the fact that I did was something of a miracle – before my schooling was further disrupted by illness, culminating in rheumatic fever. The latter ended my school days as such, and I completed my education with a 'crammer' four mornings a week. Afternoons were spent learning the practical side of both farming and estate management, something that stood me in good stead when I went to a university and gave me a far greater insight into country life than I would otherwise have gained.

2

Tuppence a Pint

T HE estate was not large, only some fifteen hundred acres. Today that would, at the most, provide work for four or possibly five people. In the twenties and thirties it gave employment to forty farm and estate workers, plus the household and garden staff who numbered another twenty. All, whatever their job, took a great pride in the estate. The many successes gained at shows by representatives from the pedigree flock and herds were a cause for satisfaction and self-congratulation. They were the mark of a job well done, whether by the herdsmen who prepared the animals, the farm workers who tilled the land and harvested the crops, or the gamekeepers who controlled the rabbit population, allowing crops to grow. It was a team effort. Each department had its head man and there was often much friendly rivalry but, in the long run, it was the good name of the estate that counted.

There were some great characters and personalities among the employees, including one Charlie Davis. He was the head carpenter, a diminutive Welshman, barely five feet tall, with a fiery temper and quite unbelievable skills. There was nothing that he could not make out of wood and it was said that anything he did would last at least a hundred years. On wet or very cold afternoons I would join Charlie and his two assistants in the workshop at Britwell and try to assimilate some of their skills. If Charlie happened to be in a good mood – and he frequently wasn't – he would regale me with stories of salmon poaching.

Charlie had been 'king of the poachers' in his homeland, and always started off any yarn with 'There was this man in Wales, see.' He never said he was that man, but father, who knew his secret, told me. On moonlight nights, when the salmon were running, face blackened, he would stand waist deep in a river for hours at a time, spearing salmon with a trident as they tried to make their way up and over weirs. At other times he and his gang would net a pool, wading into the icy water almost up to their necks. Salmon poaching was and is a very serious matter; in the early part of the century, if caught, it would almost certainly lead to a stiffish jail sentence. Charlie made things 'so hot' for himself that eventually he had to flee the country. It seemed that 'this man in Wales' was disturbed one night by a water bailiff, just as he and his mate were climbing out of a river. A running battle ensued and Charlie, realising he was dangerously close to capture, tried to ward off the bailiff with his trident. The bailiff caught it, but Charlie pushed him back against a tree and two of the prongs passed either side of his wrist. Charlie drove the weapon hard into the tree, pinioning the bailiff's arm firmly to the trunk. At that moment Charlie's mate, who had been temporarily knocked out by the bailiff, came on the scene. Quickly taking in the situation he grabbed the bailiff's other arm and 'stapled' that to the tree in like manner. The miscreants made off, having first collected the salmon they had so unlawfully taken, and headed for home. Charlie, feeling sure he'd been recognised, packed his bags and was at the station before dawn. After several years of drifting around he eventually met and married the cook from my father's home and, through that contact, arrived at Burnham.

The entire staff were terrified of him, for he thought nothing of hurling a hammer at anyone who disturbed

16

him while he was planning a job, but he was a master craftsman. Among many other things he built all the wagons and carts on the estate, even the wheels. Their rims and spokes were all laboriously cut out by hand and, when completed, the wheels would be banded, that is fitted with iron tyres, by 'Old Jay', the estate blacksmith.

Charlie suffered terribly with rheumatics, no doubt a legacy from his misspent youth standing for hours in icy rivers. Some days the pain was so great that he could hardly use a saw, or one of the big smoothing-planes. When things got too bad Charlie would strip to the waist and soak the shoulders and top of the arms of his red flannel vest in turpentine, of which there was always a good supply in the workshop. He would then wring it out, put it back on and dress, adding an extra pullover – 'to work up a good sweat, see.' He claimed it never failed. Colds also received drastic treatment. Four drops of turps on a lump of sugar every four hours! Sore throats, too, came in for a practical if somewhat primitive treatment, which never ceased to fascinate me when I was young. He would get a nice firm piece of pork or bacon fat, attach it to a length of string, swallow it and gently pull it up. He'd do this perhaps half a dozen times every hour or so. When he'd finished he'd turn to me: 'Must grease the throat, see, nothing like a bit of good pig's fat to ease the soreness.' Charlie never had a cold for long, but in spite of his success in this sphere he never seemed to make any converts to his remedies. He lived to a ripe old age and never saw a doctor his entire working life.

Charlie was a tough little man in every respect, but he had one weakness – cats! Every afternoon, around four o'clock, his feline friends, from all parts of the farmyard, gathered outside the workshop. Experience had taught them never to cross the threshold; if one tried to, a lump of wood or a tool would be hurled at it. When all were assembled, Charlie would set off for the far end of the yard to the dairy where the milk was strained and cooled. He would help himself to about a gallon and, looking like the 'pied piper', return to his domain surrounded by ten to fifteen cats. He poured the milk and stood by, smiling benignly, until it was finished.

Another great character was 'Jummy' Young. I suppose he had a christian name, but I never heard it. He was a remarkable man, who went virtually everywhere at a run. Wounded four times in the 1914–18 war, he had been shot through both hands and both feet. Four times he was sent back to the front, finally to be gassed. However, he eventually recovered, was demobbed and sent home from hospital, only to find his wife had run off with another man. Years later he found great happiness in his second marriage. Jummy believed in God, but he had his own interpretation of religion. He was adamant that there was no hell in the sense that he believed there was a heaven; this world, he often assured me, was hell and, in view of what he had gone through, who could blame him? There was nothing Jummy could not do – milking, ploughing, building a rick, laying a hedge, braiding a horse for show – all were done with equal ease and expertise. His main job was being in charge of the outside maintenance work on the estate, but if ever there was a problem it was a case of 'send for Jummy'. In addition he helped father in the garden of an evening and cared for Ralph's and my menagerie when we were away.

Apart from the usual children's pets, we had some less orthodox ones, like the two foxes, Dom and Brandy. Rescued by my brother when about two days old, they were reared with the aid of a fountain pen filler. They were lovely creatures, both males, which we had neutered when old enough. They would play happily in the garden with the dogs, one an ex-hunt terrier! Brandy was quite amenable to going for a walk on a collar and lead. Ralph once caused quite a stir by taking him on a bus into Maidenhead and walking him round the town. Dom, however, was a totally different character and, while very happy to be out in the garden with us, would rush off to his kennel if anyone strange appeared. Both Ralph and father used Brylcreem. There was nothing the cubs enjoyed more than for either of the Brylcreem Boys to sit on top of their 'den' and allow the cubs to run their teeth, like a comb, through the well-Brylcreemed hair – sheer nectar to them!

As well as the foxes we had a sow badger – Joanie. She was large, boisterous and spotlessly clean in the house. When father got settled in his armchair of a winter's evening, in front of the fire, there was nothing that she liked more than to lie on his lap. She would remain there as long as permitted, which usually wasn't very long because she weighed between 28 and 30 pounds.

Rather unusual 'pets' were three stoats. These were Ralph's, obtained through an advertisement in the *Gamekeeper* for 'Young stoats ready to eat'. Mother, when she saw the advert, wanted to know if he'd like them grilled or boiled! They were an interesting trio. The dog one could pick up and handle as though he were a ferret. One of the bitches would run over one's hand or up one's arm; she could be quietly stroked, but not picked up. The other bitch just buried her teeth in any hand that was foolishly put near her, and remained totally wild.

We had a great assortment of birds, mostly rescued, including two ravens whose parents had been killed when they were fledglings. They had been in captivity for about ten years when we gave them a home, so there was no hope of returning them to the wild. We also had three golden pheasants, a cock and two hens. Nothing remarkable about that, except that when they were nine years old the cock died. Shortly afterwards one of the hens started to develop male plumage, 'crow' like a male and chase the other hen around. This metamorphosis was monitored by a great friend of my father's, John Norman, from the South Kensington Natural History Museum. The complete sex change took between nine and ten months, at the end of which time it was impossible to see any variation from a normal male's plumage.

At one stage we were into mice in a big way and had all kinds and colours. At that time, there was a great enthusiast offering £100 for a tortoiseshell mouse. Ralph and I set out to breed one. We achieved the three colours but, alas, not in definite patches; the hairs were intermingled. By the time we reached this stage we had about three hundred mice surplus to requirements. I telephoned Gamages' pet department. 'Yes, they'd be delighted to take them at sixpence each.' We were in the money! This was the opening we had been looking for and we set out to corner the mouse market. When we had around a thousand, which was in no time at all, I telephoned Gamages again, with visions of netting £25 – a small fortune. 'Good gracious no, they'd never need mice again. The last lot were quite capable of maintaining their required numbers.' I tried all the other big London stores with pet departments. No one wanted our mice. One even said they could get all they needed from Gamages, who seemed to have a permanent surplus to requirements! Our visions of wealth receded rapidly, while our mouse population increased at an alarming rate. Finally we parted with all twelve-hundred at a penny each to the London Zoo where, sadly, they were despatched and used as food.

Perhaps time mellows one's memories, but it seemed that people, at least in the country, were more contented in the twenties and thirties than they are today. They took a greater pride in their work and had a loyalty to their employer often sadly lacking in these emancipated times. There was always a great sense of competition with neighbouring farms and estates and a strong desire to be first, whether it was sowing the spring corn, ricking the first field of hay or completing the harvest. The latter was frequently the prelude to a harvest supper. This did not happen every year on the Burnham estate, but when it did it was a slap-up do. The big barn at Britwell was emptied, swept out and decked with bunting. Even the grinding machinery was removed from the top end to make room to dance. A stage would be erected by Charlie and his helpers. Talent was not lacking, particularly after the first two or three firkins of beer had been emptied!

To begin with it was all very solemn. There was a high table, where many of the silver cups won by the stock during the year were displayed. Edward Clifton-Brown and his wife, Dolly, took the centre spot, flanked by mother and father, the architect and his wife, plus several guests. At one end would be Tom Rose, the foreman, and his wife. At the other the head gardener and his wife. An outside caterer arranged the meal, normally soup, beef or mutton with plenty of vegetables, apple tart and a good supply of freshly baked crusty bread, cheese and pickled onions. Speeches would be made, the boss's health drunk, then Mr and Mrs Clifton-Brown would depart, with their guests. Father would remain for about an hour to listen to and applaud the early entertainers. Music was supplied with the aid of a piano, fiddle and accordion – not a piano one, a real old squeezebox – on which Harry Wadman, the head shepherd, was a true virtuoso. Once father was sure that all was well, an ample supply of beer and other drinks available, he would depart, leaving Tom Rose as master of ceremonies.

While there would be a few sore heads the next day, it

was seldom anyone drank to excess; in fact I can only once remember hearing of someone passing out. At the start of the evening there were a very limited number of bottles of whisky at intervals down the tables, just enough to allow each person a reasonable tot. One delightful character, Micky Hassett from County Cork, who originally came with a gang of casual labourers in the spring to hoe roots and remained for years, when asked by father if he was alright, replied 'Be-gob sorr I am t'at, haven' I mixed th' chemicals. Good night sorr.' And with that he slid slowly under the table into complete and peaceful oblivion. Two of his mates carried him out to an empty loose-box, saw that he was well bedded down with straw and left him to recover. Late next day Micky staggered forth – it was the first time he'd tried 'the hard stuff' and, so history relates, the last.

An annual event looked forward to with great excitement was the South Bucks ploughing match. Not only was there ploughing, but also competitions for the best turned out team, best single horse – in fact, if there could be a biggest or best it was scheduled. Any of the carters could compete and as many of the staff as could be spared were given the day off to cheer on their colleagues. The horse ploughing, for many years, was dominated by Arthur Goodchild from Biddles Farm. He was such an enthusiast that he even bought his own plough – a new Ransome. A single furrow horse plough back in the twenties cost around £5 or £6 and Arthur's was cared for and polished as though it were a Rolls-Royce! Jack Keen, the head tractor driver on the estate, who lived at Lynch Hill, reigned supreme in the tractor ploughing for many years. Tom Brookling, who had been taught by Jack, normally took second place. One year there was a slight hiatus – the judges reversed the placings! It was the cause of much merriment and repeated visits to the beer tent. Everyone liked Jack, but the Brooklings were a Burnham family and Tom was a cheery and popular person.

Not long after this epic happening a pair of new cottages was completed at Lynch Hill. They were very modern: three bedrooms, sitting room, kitchen with a range that heated the water, a bath in the scullery, which was also the passage through to the back door, and leading off that was an inside lavatory. Tom Brookling was about to get married and it was readily accepted that he should be one of the lucky recipients. Shortly after Tom and his bride moved into their new home, I remember hearing a conversation between Edward Clifton-Brown and father, one Sunday morning, when the former had called at Pound Cottage on his way back from church.

Father remarked: 'Brookling is very thrilled with his new home.'

'So he damn well should be,' came the reply. '£560 for a pair of cottages. The builders must think I'm made of money.'

It sounds ridiculous, but even fifty years ago one could buy a three bedroomed house for well under £500 and a cottage for as little as £150 to £200. When Edward Clifton-Brown died in 1943 the executors, his wife and two sons, kept on the estate for several years, but in 1949/50 it was sold. Harry Jaycock, a son of 'Old Jay', the blacksmith, lived in an estate cottage in Burnham High Street. Harry had worked on the estate since leaving school in the twenties. He asked father if he might buy his home. Father enquired what Harry could go to and when told £350 replied that he did not think there would be any problem, but that he would have to put it to the executors. Imagine his consternation when told that 'Jaycock must take his chance at the auction like anyone else.' It sold for £360. Harry lost his home for £10.

There are those today who think that, in the twenties

and thirties, farm workers were only one degree above slavery, but this is a totally wrong concept. True, there was not a lot of money for luxuries, although several of those living on the estate had motor bikes and Arthur Goodchild had his plough. Holidays were rarely if ever taken, and were largely at the discretion of the employer. On the Burnham estate all had the option of two weeks' holiday or two weeks' extra pay. Almost without exception they took the pay. As far as Bank Holidays were concerned there were only four: Easter Monday, with time off being allowed on Good Friday to go to church, Whit Monday, August Bank Holiday Monday and Boxing Day. On the latter there was always a big pheasant shoot and most of the staff, other than the stockmen who had to look after their charges anyway, turned out as beaters and made an extra day's pay. A few lucky ones were tipped by the guns for carrying cartridge bags or other such services.

Midway through the twenties many started to own wireless sets. Alf Small, Charlie Davis's chief assistant, was a dab hand at making up sets. I remember he made one for father, somewhere around 1923 or 1924: a magnificent thing with a large horn speaker and a control panel something akin to that of the 'Star Ship Enterprise'. It ran off accumulators, which had to be changed and recharged two or three times a week. A 30–35 foot larch tree was felled and stripped before being erected to carry the aerial, so that we could hear 'London 2LO calling'. As the wireless grew in popularity, it helped to detract a little from the village pubs as the hub of activities and entertainment. Many went to their favourite pub, particularly in winter, for the social side and a singsong, just as much as they did for the beer. I remember hearing much grumbling from old Bill Herbert and 'Gunner' Stingmore, a veteran of the Boer War, when beer went up from a penny ha'penny to tuppence a pint! Both blamed 'they bloody wireless things', keeping men away from their beer and causing landlords to put the price up so as to be able to make an honest living.

The moving pictures came in for similar criticism and when the 'talkies' started that was altogether too much for the older generation. I remember old Gunner saying, after he had been taken to the pictures by his daughter, that such things could only lead to hooliganism and sin. Perhaps he was wiser than we gave him credit for at the time. However, I do not think the advent of the cinema affected the price of beer, for when I first legally started to drink beer, one could buy four halves of mild and get tuppence change out of a shilling!

3

Early Days with the Gun

IT was not long before I graduated from an airgun to a shotgun, a single-barrelled .410. Bob Hedges continued to be my tutor. To begin with it was a question of developing the right stance, learning to cock and uncock the gun – it was a hammer one – and to swing it effortlessly. While this was going on he drilled me in a strict safety code. Finally the day came when he put some cartridges in his pocket and we walked round to the back of Lynch Hill farmyard, where there was a rubbish dump. When we got near this he stopped, told me to move to his right, loaded the gun, pulled the hammer back so that it was cocked and started to walk forward, telling me to keep in line. Suddenly there was a bang and a substantial hole appeared in the ground to his left. He turned to me.

'I did that to show you how dangerous this is. Even if it is a small bore gun, it could blow anyone's leg off at close range.'

He gave me the gun to hold while he stood an old door up, that had been thrown on the tip. He then reloaded and fired another cartridge at it from about five yards. The result was awesome and it left me in no doubt as to the appalling results that the misuse of firearms could bring about. It was a lesson I was never to forget.

My first targets were tin cans standing on the top of fencing posts some 15–20 yards away. I was soon claiming 100 per cent success shooting these and it was decided that it was time to try a moving target. This was provided by means of a ball, about half the size of a football, rolled at speed down a steep slope. Often it would bounce over the rough terrain, making it that much harder to hit. Finally, the day came when Bob felt a 'flying' target was required. This took the form of a golden syrup tin with a few stones in it, thrown high into the air. Soon I was scoring very consistently. When I'd had nine 'kills' out of a possible ten, Bob launching the targets from out of my sight, so I had no idea whether it was going to be a 'runner' or a 'flyer', my instructor felt it time to acquaint father of my prowess.

Father was delighted, but questioned me at length regarding all safety aspects. However, Bob had not failed on this. He had been a sergeant in the infantry in the Great War and, after seeing active service at Gallipoli, was brought back to England as an instructor, before crossing to France. No new recruit could have been more strictly drilled than I had been. Any deviation from the code that had been laid down led to the immediate cessation of schooling for the day. I loved shooting and so learned quickly. My training continued and I became more and more proficient. I was anxious to move on to more exciting and difficult things than tin cans and rubber balls!

It was the day after Ralph returned for the Christmas school holiday that father told us that, subject to it being fine the next day, he was taking us ferreting and that we were both going to shoot. I would be supervised by Hedges, Ralph by father. We would both be using single-barrelled .410s and the 'dos' and 'don'ts' of ferreting were explained in detail. These we both had to repeat until father was certain that, theoretically at least, we were 100 per cent safe.

The next morning dawned crisp and bright, with just a little frost, but not sufficient to cause any problem if the ferrets became 'laid up' underground and had to be dug out. Jummy Young, who frequently accompanied father on such forays, came with us. We took three jills (bitches) and one dog ferret called Jumper. The latter would only be used if one or more of the jills, who would be worked loose, failed to reappear. This was likely to happen if they had killed underground or had one or more rabbits bottled up a dead end. Then Jumper would have a collar put on attached to a very light cord line. He would be tried down various holes until he found where the missing jill was 'laid up', and would remain there until they were both dug out. This was often a long and arduous task, but one which had to be performed. Frequently, however, it could be most rewarding, for it was not unusual to find a number of rabbits up a dead end. Years later the most I ever found, after a long hard dig, was eleven.

We met Bob at the end of Lamas Wood, where it joined Lynch Hill Lane, and made our way to the open burrows at the far end. Proudly I carried my gun, but no cartridges. These father had handed over to Bob to dole out as required. Ralph, too, carried his gun, but he was allowed to fill his jacket pocket with ammunition as he was going to load for himself. We took up our positions, Bob and I to the left of the burrow, Ralph and father to the right, and Jummy between us. Father had a gun with him, but Bob felt he had enough to do looking after me. Jummy went forward and put two ferrets to ground.

When he had returned to his place, on instructions from Bob, I loaded and cocked my gun. I was tense with excitement, but very conscious of where I could or could not shoot. I clearly heard a rabbit thumping with its hind feet, drumming out the danger signal. Seconds later, one bolted out to the right. I saw Ralph raise his gun and fire. The rabbit carried on, then there was a louder bang as father brought his twelve bore into action. The rabbit rolled over and lay still. Ralph only just had time to reload before another came hurtling out; this time he made no mistake. I was green with envy, but nothing to what I was after five more had all gone Ralph's way and I hadn't had a shot! Seven were all there seemed to be in that burrow, and both jills were out on top of the ground.

I uncocked my gun, took the cartridge out and handed it to Bob. Jummy went forward and picked up the two ferrets and put them in the box with their companions. We moved on some forty yards to the next warren. The jills were put to ground again and the adrenaline flowed as again expectations mounted. This time I was not disappointed. A rabbit came racing out of the hole nearest to me and away down the wood. I raised my gun, swung carefully on my target and pulled the trigger. The rabbit rolled over, stone dead, a clean kill, my first bolted rabbit from ferrets. I was just seven-and-a-half years old. It was to be another sixty before an unfortunate fall forced me to put my gun away for good. I added several more rabbits to my tally before we headed back to Pound Cottage for lunch. When we reached home there was a message for father – business had to take precedence over sport – and so my first day's ferreting was cut to a half-day. Although there were many hundreds to follow, it has always remained the most memorable.

Slowly and firmly my tuition continued. I spent much time, during the school holidays and on summer evenings, going around Bob's beat with him, learning about game conservation and vermin control. Normally I carried what I had come to look upon as 'my' .410 and shot a number of rabbits and vermin, but I never shot at anything until I was told that I could.

I was ten before I shot my first flying pigeon. It was a very hot August morning when I joined Bob on his rounds. As we stood outside his cottage, we could see pigeons dropping in on a piece of wheat on the far side of the big arable field below Lynch Hill Farm. Bob remarked that we might give them a call, as we set off to feed the wild ducks on Lamas Pond. It was about an hour later that we approached a big oak tree, adjoining the wheat where we had seen the pigeons going in to feed. As we neared it about a dozen flew out. A few seconds later, several hundreds rose off a section of wheat that had been flattened by a recent thunderstorm. Bob suggested that we should wait under the oak for a while and see if any of the pigeons came back.

We had only just taken up our positions when Bob whispered that there was a pigeon coming across the wheat to our left. I raised my gun as it came level – the slight movement was enough for it to see us – it jinked and started to climb at speed. I swung on it and pulled the trigger. It came tumbling down to splash into a small and sluggish stream some thirty yards away. I dropped my gun and raced across the marshy bit of ground to a plank that acted as a footbridge. The pigeon was in the centre of the stream, about two to three feet from this. Not heeding Bob's calls to 'go steady', I leaned out for my prize. With a loud splash I joined the pigeon, grasping it firmly as I surfaced! Bob was convulsed with laughter as he helped me out. On his instructions I stripped off and he wrung out my clothes. By the time I returned home for lunch I was dry.

The following Christmas holiday I was allowed for the first time to wait for pigeons coming in to roost. I was to go to Lamas Wood, Ralph to Cocksherd. He had now graduated to a sixteen bore, but still a single-barrelled hammer gun. It was a big day for us both as we set off along the footpath, outside Pound Cottage, soon after 2.30 p.m., our cartridge bags slung over our shoulders, guns under our arms. I made my way to the far end of the wood where there were about a dozen spruce, a favourite

roosting place. Close by was a patch of laurels. I constructed a hide without too much difficulty. Having got
myself reasonably well concealed, I started to scan the sky
for pigeons, the wind in my back as I knew they would
come in to land flying into the wind. It was bitterly cold.
My fingers were nearly numb as I anxiously waited.
Nothing happened and the light was rapidly failing. I
could hear Ralph shooting, and it seemed as though he
was having plenty of sport.

At last I saw three pigeons circling overhead, before
dropping in to roost. I swung carefully on the first and
fired. There was a satisfactory thump as it hit the ground.
I added four more to the bag before the light became too
bad to shoot any more. As I headed for home the snow
started. When I reached the top of Lynch Hill Lane I
stopped to see if I could see anything of Ralph coming
across the field on my right. However, it was nearly dark
and the snow, which was now coming down quite hard,
made visibility very limited. I called out a couple of times
and, on getting no answer, continued on my way. I had
been home about twenty minutes before Ralph returned,
elated. He'd shot and picked seventeen pigeons and had
spent quite a while looking for another that he could not
find. As we sat on the drawing room floor, in front of the
glowing coal fire, toasting crumpets and smothering them
with great dollops of home-made butter, we chattered
excitedly of our achievements. It seemed as though we
had reached life's zenith, for surely nothing could equal
the afternoon that had just passed?

What neither my brother nor I appreciated, at the time,
was how incredibly lucky we were to be allowed to shoot
anywhere on the estate, except in the main pheasant
coverts, and indulge our country interests to the full. The
latter, like shooting, were subject to a rigid code of con-
duct. For example Ralph was a very keen entomologist.

All butterflies and moths taken for his collection had to be correctly set when removed from the killing bottles, that is with the wings fully opened. A pin would be inserted through the thorax and the specimen attached to a groove in a cork 'setting board'. The wings would be gently opened and held in place by strips of paper, carefully pinned at either end. After a few days these could be removed and the wings would remain in place and the set specimen transferred to a drawer in a specially designed cabinet. On looking back, when one considers the very necessary stringent regulations that now govern the sale of poisons, how incredibly easy it was to obtain killing bottles. These were wide glass jars, with a cork stopper, which had about an inch of cyanide and wax in the bottom. Once a specimen was in the bottle it was dead almost immediately – they were lethal when fresh, but after a while lost their strength. When this happened they were returned to Mr Barker, the village chemist, and new ones were made up. There was no fuss or bother about this, just a question of father calling in to sign the poison book. From about the age of fourteen Ralph used to do this, and it was quite acceptable.

One aspect of entomology that always intrigued me was 'treacling' for moths. The latter are basically nocturnal and so cannot be pursued with a butterfly net to bring about their capture. One has to resort to a certain low cunning to attract them, namely the use of 'treacle'. This is an inviting concoction brewed from molasses, rum, sugar and what we used to call essence of pear drops – I forget its correct name. This having been brought to the boil and then cooled had a most inviting smell and indeed taste. I often dipped a finger in, when no one was looking, and had a lick! Just before dark we would go to our chosen place and 'treacle' patches, about 12 inches by 6, on a number of trees. It was quite a glutinous mixture, but could be easily applied with the aid of an old paint brush. Several hours after dark we would set off around our 'traps'. Frequently the 'bait' would be completely covered with a vast variety of moths, some large, some small, but all thoroughly enjoying getting pleasantly inebriated! The rum had a very obvious effect on those that consumed an excess of 'treacle' and it was not unusual to see moths prostrate at the foot of a tree, or crawling round in rather irregular circles. I went back several times soon after dawn, but always the revellers had gone, one hoped none the worse for their experience.

It was a simple matter to tap any required specimens into a killing bottle. There were always a large preponderance of Old Ladies, a variety that I have never seen in daylight, Yellow and Red Underwings and many of the smaller species, too numerous to name, were always to be seen in large quantities. Hawk moths were quite common and several varieties were regular visitors to the feast. Only once did we see a Death's Head Hawk moth, the largest of the British moths. It really was magnificent and, even in those days, quite rare. Ralph uncorked his killing bottle and then very decisively placed the cork back in the jar. It was, he said, too beautiful to kill. Alas, a vain gesture, but he was not to know of the coming wholesale carnage of our wildlife and destruction of habitat that was to be made in the name of progress.

As well as farming and shooting my great interest was ornithology. Apart from actual birdwatching, I collected eggs. Here again there were definite guidelines which had to be followed. Except in the case of vermin, never more than one egg was to be taken from a nest and only two of any species. Eggs had to be correctly blown, that is with a carefully drilled hole in the side and the contents removed with the aid of a blow pipe. These came in three sizes specially constructed for the job. This interest kept me occupied throughout the daylight hours during the latter part of March, April and May. Burnham was exceptionally good for bird life and I listed nests found, number of eggs and the number hatched for many years. Over the years I recorded quite a number of rare species, by no means least a long-tailed duck which frequented Hay Mill Pond for more than a week. The pond and adjoining reed beds covered some twelve acres and were host to a wide variety of bird life. The mill remained operational right up to and through the Second World War and the pond was a coarse fisherman's paradise.

After our first evening pigeon shooting, both Ralph and I became addicted and shot as often as we were permitted. It was not a case of going just when we felt like it, though – permission had to be sought. The guns were kept locked up and father kept the key. If the answer was 'no' that was it; there was no wheedling or pleading, parental rule over such matters being accepted without argument. In any case, we had so much to fill our days that we had difficulty in finding the time to do all that we wanted.

About two weeks after my initial pigeon shoot, I was in the estate office one Friday evening when the men came for their wages. The gamekeepers were always paid last, giving them a chance to talk to father and vice versa. I heard Bob Hedges tell father that a large number of pigeons had started to attack a piece of kale and swedes in Big Field. This was aptly named as it was 102 acres but, soon after father had taken over the management of the estate, he planted a belt of trees right across it to supply a windbreak and suitable cover for game birds to nest. Father instructed Bob to collect some extra cartridges while he was at the office and mount a counter-attack next morning. On hearing this I quickly asked if I could go too. The answer was 'yes' and it was agreed that I should meet Bob at 8.30 a.m.

There had been 2–3 inches of snow during the night, but it was a lovely bright crisp morning, with a strongish north-easterly wind, as I set off, ideal for pigeon shooting. When I reached Bob's cottage, he was waiting for me with his black labrador bitch, Judy. She was obviously enjoying the snow and it was equally obvious that she would love to have had a good romp in it. However, she walked demurely at heel as we headed up the lane past Lynch Hill farmyard. There had been no one along the lane that morning and it was easy to see where rats had been running across from the buildings to the rickyard on the opposite side. A little further on there were tracks where both rabbits and hares had crisscrossed the road. As we turned the corner heading towards Biddles Farm, we came across the unmistakeable tracks of a stoat. These continued until, some thirty yards further on, they became intermingled with those of a rabbit. We only went a few steps before we could see a confusion of tracks and several red patches in the snow. It was clear to see what had happened and where the stoat had dragged its victim into rough grass at the side of the lane. Bob slipped a couple of cartridges into his gun and walked in where the trail disappeared. Seconds later, a stoat shot out on the far side heading across the field. Bob raised his gun, fired and it rolled over, dead. Quickly he found the rabbit and hung it up on a bush, saying he'd collect it later for his ferrets.

A little further on we turned off the lane and headed across the end of a long narrow field with the unusual name of Winding Shot. This brought us to The Belt, from where we could see a steady stream of pigeons heading for the kale. Bob took stock of the situation and then said that we should walk up the centre of the trees – there was a double row of poplars, flanked with firs and shrubs on either side, giving us excellent cover right up to where the pigeons were feeding. When we got there the plan was to creep through to the edge, show ourselves and try to shoot several as they flew off to use as decoys. Bob led the way, but suddenly stopped and pointed to the ground in front of him.

'Do you know what tracks those are?' I had no idea.

'A fox, I didn't know there was one around. I'll have to keep an eye open for that fellow, for he'll do the pheasants no good at all.' There is no doubt, if one can read the signs, that snow is a wonderful means of discovering what wildlife there is in an area. A quiet walk after a fall of snow can be most revealing!

Bob held up his hand and whispered, 'You stop here. I'll go on about forty yards and, when I give the signal, creep through to the edge. Make sure you get one, no missing.'

I waited, hardly daring to breathe, for as I peered up through the foliage I could see pigeons, dozens of them, circling overhead before dropping in to feed. Bob waved me forward. As I reached the edge he gave a shout, and instantly the air became blue with pigeons, there must have been thousands of them. There was a tremendous temptation to let fly indiscriminately at such a number but, remembering what I had been told, I picked a bird and fired. Two others fell as well! I heard Bob shoot, but had no idea how many he had down. I walked up The Belt to where he stood and, just as I reached him, Judy was returning with a pigeon. She retrieved six more before we moved along for her to collect mine. Bob quickly and expertly set up the ten as decoys.

Pigeons were already returning, as I took up my position behind a bush, and I had actually shot one before Bob had found a suitable spot about forty yards away. Snow started blowing around in the wind, the sun disappeared behind heavy black clouds, it got colder and colder, but there was no time to think about that for pigeons were coming in a never-ending stream. By eleven o'clock I had run out of cartridges. (Father had given me fifty, assuring me that I would have plenty and there was no need to take any more.) The high spot of my morning was a single bird that had flown straight at Bob and, presumably when he had raised his gun, had seen him, jinked and Bob had missed. It flew along the trees towards me, climbing fast. I swung my .410 on it, fired and killed it stone dead. Easily the longest shot I had achieved, it became even more pleasing when Bob congratulated me on an excellent shot and for having 'wiped his eye'.

I made my way along to Bob's hide, where he was still shooting almost as fast as he could load. 'That's nearly it then, I've one cartridge left. Here, see how you can manage a real gun.' He handed me his twelve bore. I took it rather apprehensively – I had never fired one before and Ralph had told me that, compared with a .410, there was a terrific kick. I put the cartridge in, closed the gun and chose a target. As I pulled the trigger there was an almighty thud on my shoulder that knocked me backwards, but at the same time I saw the pigeon falling. As it hit the ground so I landed in a rather prickly bush! Bob, laughing, helped me out.

'Well that's quite a morning you've had. Your biggest bag, your longest shot and the first time you've fired a twelve bore. By the way, how many pigeons are you claiming?'

'Thirty-three including this last one.'

'Not bad with only fifty cartridges, well fifty-one if you count the last one. We'll pick what we can by hand and then I'll work Judy.'

Three-quarters of an hour later we had a heap of 95 pigeons. That made 62 for Bob, because I was definitely claiming 33! Bob said he'd seen a wounded pigeon pitch into a big elm tree in the far corner of the field, which was almost on our way home. When we reached it we found 7 dead pigeons lying on the ground, bringing our total up to 102. I enquired as to whether I could at least claim three, but was told 'no'. Birds picked like these no one claimed, unless they were absolutely certain they had shot them.

I had turned twelve when, as had become customary, a day's ferreting was organised for the first day of the Christmas holidays. Father told me he was promoting me to using his double-barrelled .410. I was ecstatic. We met Bob at the usual place at the end of Lamas Wood and I was quick to draw his attention to my new found superiority – Ralph was still using his single-barrelled sixteen bore! When I stopped chattering, Bob told father that he'd seen around thirty pigeons fly into the first narrow section of the wood. A plan was hastily agreed. I was to walk along the grass field on the bottom side. Father and Ralph would go along the plough on the other side and Jummy and Bob would walk through the wood. We were halfway along when, with a clattering of wings, two pigeons broke my way. I swung on the first, pulled the trigger and saw it start to fall. I switched to the second and this too collapsed as my second shot rang out. My first right and left! I collected my two pigeons with considerable pride and headed for the corner where there was a stile.

I could see the rest of the party standing on the ride just inside the wood waiting for me. Instructions on safety were absolute – no getting over, or through anything without first unloading the gun. There were no 'ifs' or 'buts', that was the rule. On various shooting days, I had seen some of Edward Clifton-Brown's guests getting over the very stile I was heading for; all they did was break the gun, put their thumb over the cartridges and step over. I was very proud of my achievement and equally full of my own importance. I reached the stile, broke the gun, put my thumb over the cartridges and stepped over – just as I had seen the Admiral do at the last shoot. The smile suddenly left father's face. He quietly called me to him, told me to unload my gun and give him the cartridges. He then asked for the ones I had in my pocket and finally for my cartridge bag.

Having done this he suddenly roared at me: 'Do you not know by now that you NEVER NEVER NEVER get over or through a fence without unloading your gun? Go home, clean the gun, put it away and you do not shoot again this holiday.'

And I didn't. A hard lesson but one I never forgot. Truly a case of pride coming before a fall!

4

Pin Money

IT was a warm sunny afternoon in late April as I breasted the rise in Lynch Hill Lane, just in time to see Mrs Tom Brookling leap into the air and execute an overhead smash with an ancient tennis racket! She dropped the latter, ran to the sandy bank she had been facing, leant forward, picked something off the ground and dropped it into the jam jar she was carrying. She called out to her two companions further up the lane, also both armed with old rackets: 'I've got one. The first of the season.'

What she had was a queen wasp. There was a bounty on them of 6d each up to the end of May and 3d thereafter. With extensive orchards on the estate, wasps were a real menace and anyone living on the place was encouraged to reduce their numbers. Every Friday payment would be made, once the claims had been carefully checked and counted. This started after one 'ancient' had presented a jar with a stated number of wasps in it. Father tipped them out and counted them. The number claimed was exactly 50 per cent in excess of what there were! When father pointed this out, she replied without hesitation, 'I were never good at the ole sums at school, sir.' Father passed no comment but, mixing the coins well, paid her – but 3s short. She started to turn away from the estate office window, where staff and bounty hunters came to collect their pay, when she stopped, turned to father and said: 'I'm 3s short, sir.' Father handed her the money, and commented on the fact that he thought she had said she could not count and that she had better be more careful in the future!

Wasps were not the only things that carried a price on their head. Rats were 2d each, the tips of their tails having to be produced as evidence of their death. The same price applied to young ones that had not left the nest. Moles, too, carried a bounty, as did sparrows and their eggs. The former were 6d each, the latter 6d a dozen. Sparrows were a terrible pest and it was not unusual to see a huge flock of a thousand or more rise off a field of wheat. When I was trapping around the rickyards, several of the older men would take two or three dozen home with them. They assured me that with a bit of nice fatty bacon and an onion they made a most appetising pie!

The ladies who lived at Lynch Hill were definitely top of the league when it came to wasp bashing. There was a high sandy bank which seemed to act as a magnet to the queens as they came out of hibernation. Once Mrs Cockshead, the head carter's wife, presented seventy-two as her week's offering. That earned her nearly the same as her husband's wages! There was more than a rumour going round that outside help was being enlisted. This was strengthened when one good lady in Burnham village asked Tom Rose, the foreman, to deliver a matchbox full of queens to Mrs C! Tom dealt with the matter most tactfully, in so far as he did not go to father, which would have made a major issue of it, but instead gave the lady fair warning that if it happened again the matter would be reported. From then on her weekly 'bag' was ten or a dozen – but even so, well worth having. A large loaf of bread only cost 2d, a pint of beer the same amount. It was not surprising that the Lynch Hill ladies jealously guarded what they considered to be their exclusive territory. There was no keener 'bounty hunter' than I by the time I was ten or eleven years old, but I dared not go 'wasping' on their domain.

Later in the year, any nests that were found were destroyed by 'Old Jay'. This was by the simple expedient of using a long stick to poke a piece of rag, well soaked in a strong solution of cyanide, down the hole leading to the nest. Next day he would dig out the nest and destroy the larvae. One day I came along just as he had dug out a particularly strong nest. He showed me which larvae would be future queens, explaining that they would hatch out, hibernate through the winter and appear the following spring to start their own colonies. Further enquiries brought forth the fact that the cyanide by no means always killed the grubs. For example, the nest he had just dug out had been so strong that he was only able to get his poisoned rag to just outside the entrance to the hole before he had to make a run for it. He was about to flatten the larvae with his spade, when I had a thought. I asked if I could have them for bait, as I was going fishing, to which he readily agreed. I took them off and hid them in the hayloft at Lynch Hill. Later I went back with a shoe box, made some ventilation holes with my knife, put the larvae in the box and covered it with some loose hay. Some days later 'Old Jay' asked me if the grubs were any good. I replied that they were fine and would he keep me some more? This he did and I soon had several boxes – a potential fortune. It was quite an undertaking hatching and rearing queen wasps, but I was relatively successful.

The following April I started to divest myself of my quite considerable assets. The first week I presented fifteen in a matchbox, was told well done and given 7s 6d. The weather turned very warm and the queens became very active. I lost a lot, received a nasty sting and decided the time had come to clear my remaining stock. Father was nobody's fool and a quick interrogation soon brought my secret to light. I didn't get the rollicking I expected – in fact father, while giving me a good ticking off, seemed more amused than cross. I had to pay back the previous week's money and that was the end of the matter.

Lynch Hill, quite apart from the ladies in hot pursuit of queens with their old rackets, will always be associated for me with wasps. A large dairy herd was kept at the farm and one lovely afternoon in the late summer I was bringing in the cows, from the field, for milking when one stupid animal, instead of going through the gate with its fellows, walked up the bank. She trod on the entrance to a very strong wasps' nest. They zoomed out in their hundreds, not at the cow, but straight at me! I saw them coming and fled, but alas they could fly faster than I could run. I reached the farm, literally dived through the fence and leapt into the large horse trough. I remained totally submerged for about ten minutes, except for coming up to gulp a lungful of air before hastily going under again. I was terribly stung around the neck and on the arms, and had to stay in bed for at least a week, after what had been a most unpleasant and frightening experience.

Bounty hunting was a very lucrative pastime for a small boy on a shilling a week pocket money, but it was also hard work. During the Easter holidays, in particular, it took up a large part of my time. Moles gave the best return and there was an abundance of them to trap. To begin with I had six traps. Bob showed me how to set these to the best advantage and normally I caught one or two every day. Apart from the sixpences that I collected from the office each week, the moles had an added and greater value – their skins. These were readily saleable at 1s to 1s 6d each, according to their quality. But oh how I disliked skinning them! Moles have a particularly unpleasant odour, a sickly cloying smell that seems to lodge in one's throat for hours. When skinned, I would tack the pelts out on a board, having stretched them well, and then rub in alum and salt.

Both Ralph and I were allowed to keep half of our bounty as spending money. The other half had to be 'banked' with father, until we wanted money for something he considered worthwhile. We were brought up to appreciate that money was something that had to be worked for and, when one had it, not to fritter it away on trivia. Each of us had a small cashbook, in which we had to record all lodgements and withdrawals. In spite of this I saved what I was allowed to keep to invest in my 'business', buying more mole and rat traps for, I reasoned, the more I had the more I caught and the more money I made. It was not long before I had a dozen mole traps and, on a good day, earning as much as 12s a day with tails and pelts.

My success in catching moles was greatly helped by Tubby Stockwell. He was the road sweeper who trimmed the banks and swept the roads that passed through Bob Hedges's beat. I remember being told by a friend of father's, when I was only a young lad, that there were three things I would probably never see in my life – an honest politician, a dead donkey and a road sweeper sweating through hard work. Tubby was no exception, but he was very knowledgeable in the ways of the country and an inveterate poacher. It was said that he had more tricks up his sleeves than a cartload of monkeys. One day, coming back from tending my traps, I stopped to talk to Tubby and told him of my disappointment in only catching one mole. He asked me if I was covering the 'tiller', that kept the jaws of the trap open, with soil? I replied that I was.

'Ah well, it's the scent. Give me a few days and I'll have a bottle for you.'

True to his word, the following week he presented me with a medicine bottle. It was full of a cloudy, odourless liquid, slightly brown in colour. I asked what it was made from. He smiled, spat with accuracy at a bumblebee that had landed on a poppy, chuckled and replied.

'That's my little secret. I just calls un worm oil. Just sprinkle a few drops on each "tiller" an' you'll do fine.'

It worked. The next day I had seven! I never did get the formula out of Tubby, but he supplied me with several more bottles of the concoction over the years.

Tubby, although a bit of a rogue, was a man of honour in so far as he never poached on Bob's beat. They got on well together – he'd given his word that he wouldn't touch Bob's pheasants – but he openly said he felt free to take 'an ole cock' on Bill Yeoman's beat. He was the head keeper and they had a mutual dislike of each other. Tubby was an artist with a catapult. He made his own ammunition, lead balls about the size of a small marble, and was deadly accurate to a range of up to about twenty to twenty-five yards. He once gave me a demonstration of his prowess, culminating his act by getting me to balance a matchbox on the flat of my hand, which he proceeded to shoot off with his catapult from a measured fifteen yards. The matchbox was shattered! He told me that his great joy was to make his way, on a moonlight night, through a well-stocked covert and pick off half a dozen pheasants from their roosts. He was a hard man to catch, for his weapon made no noise, and he told me he made a point of never being too greedy. His motto was get in quick, take what you want and get out. Where he operated I never knew, but there were a number of big shoots in the vicinity.

Rats were a great source of income throughout the year. In the days of corn ricks and other such havens, they increased and multiplied at an alarming rate. I regularly trapped them around the farm buildings and in the hedgerows. I must have been nearly fourteen when I bought myself a Garden Gun. It had a bolt action and was slightly smaller than a .410. It was a well-made little gun and cost about £1 or a bit less. I also purchased a small spot torch which, aided by Bob, I fixed under the barrel. After much trial and error we sighted it so that the centre of the spot coincided with the centre of the pattern of shot, at any range up to about fifteen feet.

Just after I had purchased this weapon I was at Lynch Hill one afternoon and Jim Light, the head herdsman, showed me where rats had gnawed a hole through the bottom of a door, that led from an outside yard into the feed room. He told me there had been rat droppings all over the place that morning and he was waiting to see father to arrange to have the door tinned, up to a height of about two feet. This news decided me that the time was right to try out my 'invention'.

It was very dark as I arrived at the farm soon after half-past six and made my way through the cow byre, the inmates all comfortably lying down chewing the cud. I opened the door leading into the feed room as quietly as I could and closed it with equal care. It was pitch dark, but all around me I could hear the squeaking and scuttling of rats. I could feel the back of my neck tingling, as I edged round the wall until I felt the door behind me that led into the yard. I slipped a cartridge into the breech, pushed the bolt home and switched on the spotlight. The place seemed alive with rats – huge great brutes, all scurrying for cover. One jumped, or fell, off a beam over my head, brushing my face as it landed on my shoulder before leaping to the ground, running over my foot as it made its escape through the hole to the yard. I felt myself shudder. I loathe rats and it has always required a lot of willpower to touch one. I centred the light on one sitting on a bag of meal about ten feet away and pulled the trigger: it rolled over dead. I accounted for nine before it seemed as though that was the lot. I pulled out a more powerful torch and searched the feed room, finding three that had fallen into an open corn bin. These I despatched with the aid of a two-grained fork. Finally, remembering with a shiver the one that had fallen on my shoulder, I swung the torch upwards. There, on a ledge just under the ceiling, were five more eyeing me malevolently. These quickly joined their companions – seventeen. I was about to go out to the rickyard when I heard squeaking at the top of the steps leading up into the loft. I shone the torch up to where the sound had come from and there, on the top step, were two enormous buck rats. I killed the two with one shot! I had another hunt round and, satisfied I had accounted for all the intruders, gathered up the 'bag'. With a feeling of revulsion I removed the tips of their tails and put these in a brown paper bag, leaving the carcases for Jim to dispose of in the morning.

I crossed the lane to the rickyard. There were four rows of ricks, six in each. The outer one was hay, the rest corn. I walked quietly down between the second and third line and stood still. I could hear squeaking and scuttling, as the rats made their way along the runs chewed into the sides of the ricks. I flicked on my torch and immediately had one in my sights, its eyes glowing pink. I squeezed the trigger and brought my tally up to twenty. All went quiet as I switched off the torch but, after several minutes, I heard further rustling. I switched on and again spotted a rat, blinking at me from a hole in the side of the rick. I centred the beam and pulled the trigger. Just as I did so a voice right behind me said, 'It seems to be working well.' To say that I nearly jumped out of my skin would have been the understatement of the year! I hadn't heard a sound as Bob had come up behind me, but he could move incredibly quietly – as more than one poacher had learned to his cost. Bob remained with me until I ran out of cartridges. I had only brought a box of twenty-five and hadn't had a miss so my total was twenty-nine – 4s 10d, a really good evening's work. It looked as though my mini gun was a jolly good investment.

Around this time there had been several sightings of a reddish-brown dog in the vicinity of Lamas Wood. Obviously it was living wild and was a most unwelcome visitor on the estate. Two days after my foray on the Lynch Hill rats Harry Wadman, the head shepherd, reported a ewe had been killed. They were being folded not far from Lamas Wood. When one of his assistants, Sid Goodchild, had arrived at this particular unit that morning he had seen a reddish-brown dog feeding off the carcase. As soon as it saw Sid it headed off towards the wood. What was strange about this killing was the fact that only one sheep appeared to have been attacked, contrary to the usual form when a dog or dogs kill and maim for pleasure. The three gamekeepers and George Devonshire, who was warrener-cum-gamekeeper, scoured the area for the best part of a day, but could find no trace of the dog. Arrangements were made for the keepers to sit up in turn at night to guard the unit. Two nights later another ewe was killed and partially eaten, in a fold about two miles away. Things were becoming serious. It was virtually impossible to have a guard at each unit. Forty-eight hours later there was yet another single killing, again in a different place. Nothing then happened for a week, no killing and no sight of the red dog. Everyone began to relax and it was generally believed that the culprit had moved on.

On the day it was decided that there need no longer be an armed guard on the sheep folds at night, George Devonshire was returning from a day's rabbiting. He was passing Lamas Wood, well laden with rabbits and his gear, when out rushed the red dog, growling ferociously. George had left his gun at home; he had been using purse nets. He dropped his load and rushed at the snarling animal with his spade. The latter avoided him but, far from running off, it squared up to attack again. By this time George was a good 10–15 yards from his heap of rabbits. The dog, still making short savage runs, but just avoiding the swinging spade, circled round towards the rabbits. Suddenly it broke off its attack, grabbed one and headed back into the wood. George, feeling definitely shaken, made his way to Bob's cottage. After hearing what had happened Bob collected his gun, as did George, and both men headed back to Lamas Wood but, although they searched until it was dark, they could find no trace of the marauder. Several weeks passed. Nothing more was seen of the red dog and it was assumed he had gone.

Meanwhile, I had spent several more evenings thinning out the rat population around the various rickyards. The moon was nearly full and I decided it was time to pay the Lynch Hill rats another visit. I knew it was no good trying the feed room, that had been made rat proof, but a couple of hours around the rickyard should prove fruitful. I set off down the footpath, outside my home, that crossed two fields before coming to a road, went straight over this and continued on along Lynch Hill Lane. Night held no qualms for me and the ghostly form of a barn owl, hunting up the hedgerow, was a sight to be enjoyed not feared. I headed down the hill towards Lamas Wood. A rabbit scurried across in front of me, just as a long-eared owl flapped out of an oak tree, with a startled screech. As I neared the wood it occurred to me that I had recently seen rat workings by the old pond, just past the end of the wood. Thinking I might open my account there, I put a cartridge in my gun and advanced as quietly as I could, concentrating hard for any movement on the bank that might indicate the presence of a rat.

Suddenly, right behind me, there was the most blood-curdling growl. I swung round and switched on the torch attached to the gun. There, only feet away, was the red dog! His eyes looked almost phosphorescent in the torchlight, hackles up, lips drawn back in the most evil snarl, saliva dripping from his mouth – he was a terrifying sight. He advanced one slow step. I froze. I just could not move. Then he sprang straight at the light and I felt something hit the end of the barrel. Instinctively I squeezed the trigger. With a horrendous choking, gurgling yowl, the dog dropped and slid, twitching, onto my feet. I had shot him literally through the mouth! Gasping and shaking, I moved to a nearby gate and leaned against it. Then the tears came in an uncontrollable torrent. I've no idea how long I was there before I heard a voice.

'Hello, what's the trouble?'

It was Bob and George. They had just been starting out, to patrol the pheasant coverts, when they had heard the shot. Still sobbing, I shone my torch on the dog lying in the centre of the lane. I told them what had happened.

George cut in: 'Good job, you did well. I know how you must have felt. I haven't forgotten the b . . . . . attacking me.' As he said this he picked the dog up by the hind legs and threw it over the hedge into the wood. 'I'll bury him tomorrow morning.'

I stopped shaking and blew my nose vigorously.

'You did well to wait, with a pea-shooter like that, until he was so close,' said Bob, putting his hand on my shoulder. I didn't tell him I had been so petrified with fear that I couldn't move, but I was beginning to feel better. I managed to laugh and mumbled something about it all being a matter of timing.

I accompanied the two men back up the hill. When we reached Bob's cottage he suggested we all go in for a cup of cocoa. Half an hour later Bob and George set off on their rounds, and I headed for the rickyard feeling much recovered. I had reasonable sport, well into double figures, before I decided I'd had enough and headed for home. As I approached the end of Lamas Wood I slipped a cartridge into my gun and switched on my big torch. I hesitated then, gritted my teeth, sprinted past the wood and up the hill. As I passed the big oak, the long-eared owl, who had returned, flapped out with a screech. For a fraction of a second my heart stood still. I whipped round, the gun coming to my shoulder, but there was only one red dog and he was dead.

When I was safely back in the house I nonchalantly told father of my escapade, leaving out the fact that I had been numb with fear. He congratulated me and said he was sure 'the boss' would consider I had earned the reward. I hadn't known there was one. Apparently, Edward Clifton-Brown had become so incensed over the sheep killing that he had told father to inform the gamekeepers there was a bonus of a fiver for whoever shot the red dog. Three days later I received four crisp one pound notes and two equally new ten shilling ones. I asked father to 'bank' the lot – I was saving up for a shotgun – a double-barrelled twenty bore!

5

A Spotted Pig and Garnet

THE main projects on the estate were the shoot and the registered flock of Hampshire Down sheep. The two were in many ways complementary to each other, because a large acreage of root crops had to be grown for the sheep which, in turn, provided excellent cover for game. To say that these were the main ventures is in no way to decry the achievements of the herds of Berkshire, Tamworth and Large White pigs maintained at Biddles under the able care of Fred Goodchild, the head pigman. Representatives from the two former herds were exported to all parts of the world and it is sad that these two breeds must now be classified as 'rare'. In the twenties and thirties they were much sought after as a sound economic investment, producing top quality porkers and excellent baconers. The Berkshires were particularly attractive, being black with a white blaze down the face, white feet and a white tip to their tails. Berkshire/Tamworth crosses were sandy with black spots and were much in demand for bacon. They also made excellent pets. I know, I had one!

Although Britwell was not the centre of the pig unit there were four large styes, a common feature in any farmyard built during the last century or early in this one. Fred Simpson was an excellent all-round stockman and so usually there were four sows with litters at the farm for him to care for, nearly always crossbreds. Normally a runt in a good sized litter would be given 'a tap on the head', for they would seldom thrive and invariably got pushed out from 'the milk bar', remaining unhappy miserable little things that eventually died. One morning I arrived at Britwell just in time to see Fred about to give the *coup de grâce* to the runt of a Berkshire/Tamworth cross litter, a little gilt, a minute creature only about one-quarter the size of her brothers and sisters. I shouted to him to stop and, running over to the styes, took her from him and said I would give her a home. He seemed uncertain. I moved hastily away, holding her close to my chest, in case he tried to take her back. To show her appreciation she wet all down the front of my clean pullover!

Fred smiled. 'Oh well, but don't blame me if you have to bring her back. She has suckled and had some colostrum, so should be all right to bottle feed.'

She was a pathetic little mite, but with a certain rakish look, for she had a black patch around one eye. She grunted feebly as I carried her home and put her in a rabbit hutch. I had given her a quarter of a baby's feeding bottle of milk before anyone knew she was there. I named her Teresa and from that first feed onwards she never looked back. Within a few weeks she was requiring larger accommodation. When allowed she was my constant companion. There was nothing she loved more than to follow me around the garden paths, grunting contentedly as she trotted along behind. The majority of our pets had to intermingle and it was not long before Teresa was coming for a walk with the dogs along the footpath that ran past the house and across the fields. When they had a run she would run too, giving ecstatic little squeals and kicking up her hind legs. When the dogs walked at heel, Teresa walked at heel!

Her idea of sheer bliss, when small, was to lie in the crook of my arm and have her tummy tickled, eyes tightly closed, lips drawn back in a definite smile. As she grew, so, through sheer weight, this became impossible. However, when she felt in need of a good tickle-cum-scratch she would suddenly collapse, wriggle to expose as much tummy as possible and give what we called her tickling grunt, in the hope that someone would indulge her. Once one of us got busy scratching her tummy gently with a stick, the grunt would change to little short guttural noises of uninhibited delight. She grew and grew. Jummy and I constructed a temporary stye for her, but in spite of her ever-increasing maturity she was still a child at heart and loved her walks with the dogs.

One particularly amusing incident comes to mind, when I was returning to Pound Cottage after an evening stroll and had let the dogs run on as usual for the last hundred yards or so to the back gate, where they would wait. Teresa, invariably, came gambolling and grunting along behind. There was a slight bend in the path and momentarily the dogs were out of sight. On the evening in question they disappeared from view as usual but, almost immediately, the peace of the countryside was shattered by the most piercing screams. I rushed forward to be confronted by an elderly lady, crowned with a straw hat bedecked with roses, one hand clasping her skirt firmly to her knees, while the other waved a walking stick over the now prostrate Teresa, who was busy wriggling into the best tickling position and grunting expectantly. The poor lady, thinking she was being attacked by a mad pig, had brandished her stick. This to Teresa meant a tummy scratching session and she had obligingly collapsed at the nice lady's feet! I explained the reason for a now inert, but expectant pig lying across the footpath, but the lady took a lot of convincing. Eventually she gave Teresa a little scratch with her stick and, realising that was that, Teresa got to her feet and walked happily behind the lady and me to the gate where the dogs were waiting patiently. As we stopped and I again apologised for Teresa's behaviour, the lady gave my spotted friend a little rub with the end of her stick. Teresa, with a happy grunt, advanced upon her. The latter gave one look and fled. I subsequently learned that she had moved into a house only a short distance from the start of the footpath but, strangely, I never saw her use it again!

Not many weeks after this event father came into breakfast one morning looking very serious. 'I've been looking at Teresa. She's too large to remain here and anyway she's heavy enough to go to the butcher.' For a moment there was a stunned silence. Then mother attacked. She had a great love of pigs and had once rescued a Berkshire runt which had become a much-loved pet – and she was very fond of Teresa. Murderer, callous brute, heartless wretch were but a few of the barbs that rained down upon poor father during the tearful and heated argument that followed. Eventually a compromise was reached. Teresa would go to Britwell and remain there as a brood sow. This was a tremendous concession, for all the breeding stock on the estate were pedigree and 'bred in the purple' and here was Teresa, sandy with black spots, an obvious crossbred, joining the elite.

When the time came I moved her to Britwell by the simple expedient of letting her trot along the road with the dogs and, trustingly, she followed me into the stye. On finding she was being left she voiced her disapproval in no uncertain manner. Her squeals were deafening! For some weeks after this every time I went to Britwell and she heard my voice she would start to create. She would run around her stye giving high pitched grunts and squeals until I opened the door and took her for a walk around the rickyard at the back of the barn. She grew enormous, bearing out a remark that Harry Wadman, the head shepherd, had once made to me when I commented that a newborn lamb was very small: 'Ah, her's got all the world to grow in.' Teresa was living proof of his wisdom. In due course she was mated to a Large White boar and she produced and reared, with a little help, seventeen piglets. She became a staid matron from then on, but still enjoyed a walk if given the opportunity. She lived to a ripe old age, breeding eleven litters totalling 101 piglets!

It was not long after Teresa moved to Britwell that Garnet came into my life. No, she was not a red precious stone, although she was certainly to become precious as far as I was concerned. She was black, had four cloven hoofs, a modicum of devilment to go with them, a cute little bovine face and a loud moo. She was, in fact, a Dexter heifer with the registered name of Whitewyck Garnet. Dexters are the smallest breed of cattle; originating in the west of Ireland, they stand only thirty inches or less at the shoulder when fully grown.

Harry Hobson, senior partner of the firm of auctioneers of that name, specialising in pedigree stock, was a great friend of father's and had for some time promised Ralph and me a Dexter heifer. (This was no great thrill for my brother as he was not cattle orientated like me!) One cold foggy evening in November, just before 5.00 p.m., a telegram was delivered to Pound Cottage; it was addressed to 'Twist brothers' and it was to inform us that a Dexter heifer had been despatched in the guard's van of a train arriving at Taplow station at 5.22 p.m. Taplow was the next station to Burnham, but was larger and had a far more regular service.

In this modern age it must be hard, for the less old among us, to visualise a heifer, however small, travelling in a guard's van on a passenger train! For that matter it must be quite hard even to visualise a guard's van, a large compartment where, when going on a journey, one could 'park' one's heavy luggage, which would then be carried on one's arrival at the station by a helpful and smiling porter. A wide assortment of goods moved around the country by this method – small animals as well as parcels, and even unattended children! I well remember visiting my ex-nanny in Oxfordshire when I was about six. I was labelled, placed in charge of the guard, who would have been suitably rewarded, and had a most pleasant journey in the van playing with two terrier puppies en route for Banbury. I was returned home in like manner.

After receiving the telegram there was not a moment to lose for, in those days, trains ran on time and the Great Western Railway took a great pride in the service they gave. Father got the car out of the garage, an Austin 12, and spread two sacks on the floor between the front and rear seats. We set off for the station, some 7–8 minutes away. The train drew in, puffing vigorously and dead on time. One of the porters accompanied us to the van, had a quick word with the guard, who was standing on the platform, and lifted Garnet out. She was just ten months old and stood about two feet at the shoulder. She trotted quite happily along the platform on her halter, much to the amusement of both passengers and staff. I climbed into the back of the car and Garnet was lifted in. Totally engrossed in my new acquisition, I sat in the back seat holding her, and although she had been equally restrained during her journey from Berkshire, we arrived at Britwell without mishap!

Garnet was escorted into a large loose-box by Fred and me and given a feed. She seemed in no way overawed by her journey, nor by her change of environment. In fact she immediately made herself at home and suddenly started to frisk around her well-strawed quarters, finally giving a diminutive bellow and sliding into me, catching me behind the knees so that I landed in a heap with her on the floor. I hugged her. She was mine, well Ralph's and mine, but I knew he was not interested. He hadn't even come to the station! As I sat on the floor, I thought of the fun we'd have together. I wasn't to be disappointed. In due course we received her pedigree: she was truly bred in the purple, it was as good as one could get in the breed. Her sire was twice Champion at the Royal Show. Her dam was a prolific milker and also a Royal Champion. Later we were to learn, from the report on the dispersal sale of the herd from which she had come, that she had cost 13 guineas – a lot of money in those days!

While ultimately she would become a staid dairy cow, Garnet was, in her early days, very much a pet, another companion that became part of my everyday life. Every morning at 7.15 I would leave the house, jump on my bike and race off to Britwell, all of three minutes away! I had a little bulb horn, like an old fashioned motor horn, on my cycle. As I rounded the bend, just before reaching the farm I would blow it vigorously. It was not long before Garnet realised it heralded my arrival and an equally short time following this that she decided that acknowledgement was necessary. The moment I gave a peep on my horn, she would answer with a series of well-sustained moos until I went to her.

I quickly found out that she had a reasonably high bovine IQ and it was not long before she learned several tricks. She would go down on one knee on command and remain there until suitably rewarded with a few dairy nuts. She would also stand with her two front feet on her upturned feeding box. (To begin with she was too small to reach the built-in manger and so a box was required.) It was easy to teach her to do this. All I had to do was to turn the box over, stand on the opposite side to her and hold something she liked to eat just out of her reach. She quickly discovered that she gained extra height from the box and it was not long before she learned to turn it over herself. She would then mount it and wait until she received a reward. What she really enjoyed was playing hide and seek. Someone had to restrain her while I ran off and hid – it had to be somewhere fairly obvious, like sitting up in a wagon, or halfway up a ladder leaning against a rick. She would rush around until she spotted me and, when she did, she positively bellowed with excitement and, at times, would leap into the air, all four legs off the ground at once. She would keep this up until I went to her and gave her a prize – she always had a titbit when she'd been a clever girl!

It was inevitable that she would go for walks with the dogs. To begin with she wasn't much bigger than a labrador retriever. She loved to gallop with them in the fields and would playfully butt at them, but never once did she touch one. She seemed very much to take her time from the dogs, for when they ran back to us Garnet would, invariably, come too.

By the time Garnet came on the scene both Ralph and I were allowed to take a gun out and go off around the estate on our own. One evening in May, when Garnet was about sixteen months old, I had her out for a walk and met Ralph, accompanied by his Scottie, Wendy. They were off to shoot rabbits, so I joined them. We went into a field of rough grass and it was not many minutes before Wendy had hunted up a rabbit. Ralph shot it. Garnet chucked up her head, but made no attempt to run off. A few minutes later Ralph added another to the bag. I 'hocked' them and hung them across Garnet's back. She made no objection and on we went. Several more were added to her load before Ralph shot a rabbit in the head. As so often happens with such shots, the body jumped around where it had died. This was too much for Garnet. With a high pitched bellow she rushed forward, jumped up and down on the dead rabbit and pulverised it, shedding her load as she did so. We could never break her of the habit, although we took her shooting many times after this initial outing and she became a most useful 'game cart'. In fact she became worse and I had to restrain her, on such forays, by keeping a halter on her.

By the time Garnet was two years old she had graduated to joining the Red Poll herd when turned out to graze. The cow byre, for those days, was very modern with tubular steel divisions and the feeding manger sectioned so that the cows could not steal each other's food. In the afternoon, before the cows were brought in from the pasture for milking, the basic maintenance ration was put out for each animal and dairy nuts, which by this time were available, added according to each individual's milk production. The moment the field gate was opened Garnet would be off at the gallop. She almost skidded into the byre and would run along under the stall divisions, for the full length of the shed, grabbing a mouthful of nuts wherever she could before taking up her position in the end stall.

The time came when her fun days were over; she had grown up. She was taken off and mated to one of the leading bulls in the breed. As soon as this happened she ceased to be a bovine extrovert. No longer did she want to play hide and seek, although she still enjoyed a run with the dogs. In due time she produced a bull calf. Alas, this was always to be the case and I was never to have the heifer that I could run on as an eventual replacement for Garnet. However, when she calved the action really started, for she was an absolute little tiger to milk. Garnet suddenly made it very clear that she did not wish for a life of domesticity. She kicked, bucked, plunged, in fact did everything she could to avoid being milked. It took three men to restrain my pint-sized cow, even with a kicking strap holding her hind legs together, while Fred milked her. This happened night and morning for several weeks.

By this time I was an experienced and competent milker, but Fred would not let me try. However one afternoon, just before milking was due to start, I found myself alone in the byre. I picked up a bucket and stool and moved in beside Garnet, uttering soothing noises as I did so. I sat down and, still talking, started to milk her -- she never moved. It was holiday time and for the next two weeks I milked her night and morning. She was never any more trouble and, like Teresa, lived to a ripe old age.

6

Pheasants and Gundogs
with a Difference

INCUBATORS, while widely established for hatching hens' eggs, were not considered reliable enough, in the twenties and thirties, for normal use in producing pheasant chicks. These were still hatched under broody hens and here the activities at Leas Farm dovetailed in with the shoot to a small degree, for the farm was entirely given over to poultry. Battery hens and broiler chicken were still a thing of the future. Equally, it was not a free range system in its true sense that was practised. Acres were covered with wire netting pens 50 yards by 50 yards, laid out with precision. Between them were 'roads' wide enough for a horse and cart to pass through. Each pen had a house to accommodate a hundred birds. There were smaller pens for breeding units and poultry largely for exhibition. The main breeds were Rhode Island Reds, White Wyandottes and White Leghorns, the first two supplying the majority of broodies required by the shoot. The Leghorns were not, on the whole, good mothers and were only considered for maternal duties at times of real shortage, when nothing else was available.

Those broodies that failed to do service with the game-keepers, or were not required for hatching sittings of hens' eggs from some of the less common breeds kept, suffered the nearest thing to intensive housing that had so far been devised. All along the north wall of one of the barns was a long coop, some four feet off the ground, the ends, front and floor being made with slats. The roof was constructed so that rain, and indeed snow, early in the year, was kept off the inmates. Hens that became broody and were not wanted for sitting went into what Cyril Barker, the head poultryman, called the 'freezer' coop – the floor being slatted, there was a continuous flow of cold air under maternally minded hens! A good north-easterly breeze was certain to dispel any such desires and the majority would be back with their companions within three to four days.

The main pheasant hatching unit was at Bill Yeoman's home, Swilly Farm, although the meagre quantity of buildings hardly justified the title. However, there was more than enough room for hatching out the pheasant chicks. After the end of the shooting season, the required number of hen pheasants would be caught up. This was done by the use of traps: cages, baited with corn, into which the pheasants could walk, but they were unable to find a way out. Of course the requisite number of cocks had to be caught as well. When taken to the breeding pen the flight feathers, on one wing, were clipped so they could not fly out. Fir branches would be cut and arranged in the pen to provide nesting sites. Once a number of hens were laying, the eggs would be collected each day. When there were sufficient, that is several hundred, they would be put in the nest boxes under broody hens. The latter would already have been sitting, for several days, on dummy eggs to settle them down. Outside the hatching shed, on the grass, were lines of sticks driven into the ground, each with a string attached. Every morning the hens would be lifted off the eggs and taken outside, a string attached to a leg, where they had about 8–10 minutes to eat and drink before going back on the nest for another 24 hours.

Meanwhile the rearing field would be prepared. Strips would be mown across the meadow and coops set out about ten yards apart. Between the rows the grass would remain uncut, so as to provide cover for the poults as they grew older and ventured further from home. When the eggs hatched, hens and chicks would be transported to the rearing field. The hens would be put in the coops, each accompanied by twenty chicks. This meant that, sometimes, there would be a surplus of foster mothers, if there had not been too good a hatch. These were returned to Leas Farm, if they had come from there; if not they joined Bill Yeoman's fairly extensive free range flock. Even with all the hens on the estate, it was often necessary to buy in broodies. They usually cost anything from 3s to 5s and many ended up as boiled chicken with parsley sauce on Yeoman's kitchen table! He had a large garden and a good supply of parsley!!

For the first few days the pheasant chicks were fed on hard boiled eggs. There was a mobile hut, with a coal-fired stove in it, on the rearing field, which was used as a shelter-cum-canteen by the keepers. Adjoining this each year a lean-to was erected with corrugated iron sheeting. This housed, among other things, a large copper in which vast quantities of eggs would be boiled. These were obtained from hatcheries, being clear eggs removed from incubators at ten days. (It is possible at this stage of incubation, with a strong light behind it, to see if an egg is clear or whether it has an embryo in it.) Such eggs could be bought for tuppence a dozen. While they were cheap food they were certainly labour intensive, and I spent many hours on the rearing field as a boy – a large number, in fact, peeling hard boiled eggs! When peeled they all had to be mashed, fine enough for the chicks to eat. They graduated from egg to very finely kibbled grain, which became larger as the chicks grew. Finally they had whole grain as poults, just prior to being moved to the woods.

Every evening at dusk the keepers had to go round the

coops, shutting them up for the night, with the solid front, which safeguarded the inmates against predators. At dawn they went round again to let out the chicks and each morning counted the number that came out from one-third of the coops. These figures were carefully recorded and should there be a marked discrepancy it was an indication that chicks were being taken by marauding vermin and steps would be taken to rectify the problem. Vermin included cats, the bane of every gamekeeper's life, and they were particularly bad at Burnham. It was not at all unusual for people to drive down from London and abandon both unwanted dogs and cats in Burnham Beeches. Being a beech forest there was little cover, and consequently no ground game on which they could feed. The result was that the poor creatures wandered onto the estate where, in an effort to survive, they became vermin.

When I grew older I frequently helped shut up at dusk, before going back to the keeper's hut for a brew and a chat, often staying long after dark if we happened to cook some kippers. Cases of these were supplied to both the gamekeepers and shepherds when they had to be up all night, either on the rearing field or for lambing. One night I was sitting in the hut with George Devonshire, who was on duty, having just demolished a couple of large kippers and a mug of tea, when I suddenly heard a noise, like wind rustling grass. But it couldn't be that, for it was a very still hot June night.

'What's that?'

George listened. 'Only the wind in the grass.'

I stood at the half-door of the hut. 'It can't be. It's only coming from the one place on the hill and the grass just out here isn't moving.'

George joined me at the door. 'Must be the wind, it can't be anything else.'

After a few minutes I said, 'That's no ordinary wind.'

George listened again. It was getting closer. 'You're right. What the devil can it be?'

I didn't know, but it was spooky. I could feel tingling on the back of my neck. The rustling was coming straight towards us and getting louder. Then I heard squeaking, reminiscent of the rickyard at Lynch Hill. I grabbed the big torch with a spot beam and shone it where the noise was coming from, some fifty yards away. I gasped, as it revealed hundreds of pairs of eyes glowing pink in the torch light. Rats!! Literally hundreds and hundreds of them and coming straight for the hut! I'd heard of armies of rats, but had never really believed in them. George grabbed his gun, then put it down. He too had heard horrific stories of what had happened to people who had tried to kill rats, when on the march, as these were. I switched off the torch. The only light was coming from the hurricane lantern that hung from the ceiling. We stood, not saying a word, as the squeaking, rustling procession passed either side and under the hut, to disappear into the beech wood behind. We did not say a word until we could no longer hear anything. What George then said is certainly not repeatable. It had been a hair-raising experience.

George opened the door and stepped out, then reached in for his gun.

'Come on, let's go and see what damage they've done.'

I picked up the torch. It was easy to see the route the rats had taken across the field. The grass was flat. They had passed either side of a number of coops, but had not attacked anything. All was quiet. We followed their path out to the road, then back to the hut and beyond. There was a clear and deep track through the beech leaves in the wood.

I arrived home just after midnight. Father was waiting up for me, and he was not pleased. Further, it was obvious that he did not believe me. With a curt 'I'll talk to you in the morning' he departed to bed. Next morning, when I went downstairs, he was waiting for me.

'Come on, we'll go and see where this army of rats that you said you saw went to.'

'It wasn't only me. George saw them too.'

We drove in silence to the rearing field. George, who should have gone off duty at 6.00 a.m., was still there although it had just turned 7.30. He and Bob were talking as we got out of the car.

'Morning. What's all this nonsense about an army of rats?'

'It's not nonsense, sir. Look up the field, you can see where they flattened the grass, and look here.' George led the way to the edge of the wood behind the hut. A track about twelve feet wide was clearly visible in the beech leaves and mould beneath the trees.

'Well I'm damned. Have you followed their trail?' Father looked at George, but it was Bob who answered.

'Yes, we followed it to the spinney on the far side of Front Meadow and it just vanished. The trouble is, with all the briars under the trees it's almost impossible to see the ground, let alone any tracks.' Father stood silently for a minute and then turned to George.

'How many do you think there were?'

'I wouldn't like to say. Many hundreds, but I didn't look too closely. I don't scare easily, but they gave me the creeps.'

Father put his hand on my shoulder. 'Hmm, army of rats, eh! Come on, let's go home, or we'll be late for breakfast.' As we walked away I winked at George. I'd told him the night before that the old man wouldn't believe me.

Once the poults had sufficient feathers to enable them to get up into the lower branches of trees, or into bushes, they were moved into the coverts. This was quite a job. Early in the morning the coops, containing the foster hens and poults, would be eased onto a piece of heavy sacking or canvas and firmly secured. This had to be done with great care, so as not to break any of the poults' legs. Then the coops, with their temporary bottoms, would be loaded onto one of the waiting horse-drawn four-wheeled waggons and taken to the woods. There they would be unloaded and the coops set out along the feed rides. When all were in place and everything was quiet, the poults would be let out and fed. They soon became acclimatised to their new surroundings and, once this happened, it would not be long before they would cease to go into the coops at night, preferring to flutter up to roost as nature had intended.

The hens would then be returned to the farm, or find a place in Bill's flock. The coops would be collected and creosoted inside and out, before being stored for the next rearing season. From then on it was a dawn to dusk job for the gamekeepers, feeding, protecting their charges against predators and, once the corn was cut, 'dogging in'. Pheasants are peculiar creatures. It does not matter how well they are fed, they always want to roam, and so it was a daily chore, in certain places often several times a day, for the keepers to go round the boundaries driving their charges back to the coverts with dogs.

Burnham was not a big shoot. The main woods were only shot three times: the second Saturday in November and December and finally on Boxing Day. Normally a bag of between 380 and 500 would be shot on the first day. Only once do I remember this being exceeded in any great numbers and, when it was, Edward Clifton-Brown was not a happy man! Normally there were eight in the party, each with a loader and shooting with two guns. This particular year a young American banker had been invited, whose name was Gibbs. He had brought his own man over from the States to load and drive him around. He was indeed a yuppie of the early thirties! He was the best shot I ever saw and Edward Clifton-Brown was ranked in the top ten pheasant shots in the country. Father could outshoot him two days out of three, so I was used to seeing good shooting. At the end of the day, according to Mr Gibbs's loader, he had fired 502 cartridges for 501 pheasants. I watched him on several drives. He never took a low bird and, if there was a choice, he always went for a cock rather than a hen. The other seven guns were all excellent shots too and had their fair share. The game larder had never been so full, with over 900 head in it. This made quite a hole in the stock of 2,000 put down in the main woods.

Apart from the three major days there was a shoot almost every Saturday. These would be mixed bags of partridges and pheasants, normally anything between 50 and 80 head. Real fun days, where quality was of far greater importance than quantity and shooting stopped throughout the season by 3.30 p.m. There was none of this shooting until dark, as is so prevalent in the commercialised shooting of today. The birds were left in peace to feed, before going up to roost.

It was on such a day, cold and with snow blowing in the wind that, for the only time in my life, I saw a labrador retriever under the influence of drink! It was Edward Clifton-Brown's own dog – Smoke. The guns had lunched at Biddles Farm House, where Tom Rose lived, but, as at Leas Farm, a room was retained for lunching in on shooting days. At these the camaraderie ran high and there was much laughing and leg pulling. E.C.B. as the owner was frequently referred to, had a great love for Drambuie and it was customary for a round of this to be served before the party finally ventured forth for the afternoon's sport. According to father, who was present, Laurence, the butler, spilt some on the floor. Smoke, who was nearby, quickly licked it up with much gusto. This led to a certain

69

amount of ribbing before Colonel Warren, a delightful Irishman and Chief Constable of the county, told E.C.B. not to be so mean and to give his dog a decent drink. Laurence was sent for a saucer and a generous measure of Drambuie poured. Smoke lapped it up with relish and, when I met the party, as they came out to start the afternoon, he seemed none the worse.

The guns made their way to the pegs for the next drive and E.C.B. suggested I should stand with him. Smoke sat down in front of his master, apparently quite sober and in every way normal, for he'd even managed to cock his leg at a nearby tree without becoming unbalanced. However, it was by now freezing hard and, as is well known, cold air can do strange things to those who 'have drink taken'. Behind where we stood was stubble, with a light covering of snow. Not a lot came our way, but Edward Clifton-Brown accounted for a right and left of partridges and a hare. The drive was over and Smoke was sent for one of the partridges, which could plainly be seen. He went out at great speed, but hardly in what one could have described as a straight line! Although he tried to pick it up, he overran the partridge by some 8–10 yards. He had another go and again failed to make contact. The third try he slowly stalked it before lying down and, apparently, tried to focus on his retrieve, before crawling on his stomach for the last 2 or 3 yards. Finally he had the partridge in his mouth and, legs wavering, delivered it to his master!

Smoke knew there was more work to be done and, without being told, set off again. This time the hare was his objective. Normally he would whip-up a hare and gallop back with it, for he was a big strong dog. This time he made his approach in low gear, but then the trouble started. When he lifted the hare, albeit with difficulty, it seemed that his legs would not work. Finally, after a number of abortive efforts to bring it back, he just sat down holding the hare, looking dejected and whimpering quietly. E.C.B. went to his rescue, took the hare and instructed Bobby Yeoman, Bill's son, to take Smoke back to The Grove. Clifton-Brown then turned to his guests and said, quite solemnly, 'You know, I'd never have given it to him if I hadn't thought the silly ass could hold his drink.' I'm glad to say that Smoke was none the worse next day and remained strictly teetotal from then on.

While this had been going on father had been watching, his gundog sitting quietly beside him, partially camouflaged by the snow, for she was small and largely white. She was in fact Bridget, a fox terrier with great emphasis on the fox as she had been a working hunt terrier. Father delighted in having something different and, apart from the traditional breeds of gundogs, frequently brought out ones that could hardly be said to be orthodox, but were always most effective. Bridget was just such a one. Some years previously father had been visiting his old stamping grounds in Oxfordshire and had called at the Heythrop Hunt kennels, with which pack he'd hunted before moving to Burnham. As he was leaving he saw, curled up asleep on a sack of meal, a very attractive terrier bitch. Laughingly he had said, 'I'll take her with me.'

To his surprise the huntsman replied: 'Do by all means. There's only another three weeks to the end of the season and she'll be put down then. She'll be four before the start of the next one and we don't keep any terriers that old.'

So Bridget came to Burnham and quickly became father's 'shadow'. He seemed to have an affinity with all animals, and dogs in particular. He soon discovered that, as he described it, Bridget was 'ginger' with rats. Also she loved to hunt, but never seemed to go more than 25–30 yards from him. One afternoon he took a gun and hunted up some rough grass and briars with Bridget. It wasn't many minutes before she put up a rabbit, and father shot it. Much to his surprise Bridget went straight to it and, without hesitation, picked it up and brought it to him. He carefully examined the rabbit and found that she had not damaged it, which he felt was nothing short of miraculous in view of the way she would smash a rat.

Not many weeks later father was again passing the Heythrop kennels. He called in to say what a great success Bridget was proving to be and commented on how extraordinary it was the way she retrieved. The huntsman laughed.

'That was young Billy who taught her that trick, one of my kennel lads. The little so-and-so used to take her up to the golf course on a Sunday morning. There's a spot where the players drive over a rise. The fairway is narrow and balls are always landing in the rough. Billy to begin with had Bridget retrieving off the fairway, but after a while she became really good at finding them in the thickest cover. Billy used to help look for balls and when he'd "found" one he'd give it to the owner. It nearly always resulted in a tip. He'd often come back with five bob or more.'

That conversation gave father food for thought and he set about training Bridget properly. She was a most apt pupil. She did not appear in the shooting field the following season, but the next one saw her there – rock steady, sitting out in front of father at a pheasant or partridge drive, as bright and bold as any pukka gundog. To begin with, both she and father were the butt of much friendly witticism. However, Bridget had a quite extraordinary nose and it was not long before the witty remarks were being replaced by ones of astonishment and admiration, as she quickly gained a reputation for being a very reliable game-finder.

I well remember one Boxing Day shoot. The last bird down was a very strong runner, which quickly disappeared into a nearby spinney. E.C.B. sent Smoke for it, but without success. Several other labradors were tried. Light was fading fast and it was getting colder by the minute.

One of the guests turned to his host and said, 'Come on Ted, it's not going to be picked. If my Grouse couldn't, nothing will.'

Clifton-Brown turned to father: 'Are you going to let Bridget have a go?'

Father sent out his little bitch, who had been patiently waiting. Grouse's owner gave a positive snort of derision – but he did not know Bridget. We waited about ten minutes, which is an awfully long time under such circumstances, before she reappeared bearing a very live cock pheasant with a broken wing! Micky Hassett and another beater, going home, had seen her coming back across a road, with the pheasant, over a quarter of a mile from where we waited.

Years later, after Bridget had moved on to those halcyon coverts in the sky, father had another 'odd' gundog, a bull terrier called Sally. She was a very useful game-finder but, try as he would, father could not stop her running-in to a fall. She was a truly excellent marker, really spot on, and would hunt close to one for hours – a really good rough shooting dog. Her nose was not as good as Bridget's, but on a good scenting day she could take a runner with the best of them. Like Bridget she was a ratter, but had not the skills of the older dog and not infrequently was 'nipped' by her victim before she could finally despatch it.

Sally had several weaknesses, the most unusual being for the icing on cakes. She was not by any means a habitual thief – in fact one could leave joints of meat within her reach and she would ignore them, as she would other foods. It was just cakes! Twice she lifted a cake on a plate off the table and nibbled and licked off all the icing, leaving the remainder for us.

7

Hay and Harvest

————◆——————◆◇◆——————◆————

THE year's work on the farms culminated in haymaking, followed quickly by the harvest – but how different things were in the twenties and thirties compared with today. Now vast machines swallow up the crops like some monster from the Neolithic age! But then there was almost an air of excitement as meadows came to fruition, or a one-year ley of red clover, in an arable rotation, burst into flower, indicating that the time had come for mowing. In the weeks preceding, machinery would be given a final check, greased and oiled where necessary, blades sharpened. When the weather looked settled, mowing would start. In those days weather forecasting was in its infancy and had not become the science it is today. Traditional and tried methods of foretelling the weather had to be relied on, but, like today, could prove to be wrong. Swallows and house martins were good indicators of the immediate weather. If they were flying high, then there was an excellent chance that it would remain fine. However, if they were just skimming the ground, then that foretold rain. Seldom were our feathered forecasters wrong.

The mowers in the twenties were nearly all horse drawn; just a few were pulled by tractors, but they were the same machines with the long poles by which the horses drew them shortened to a draw-bar to attach them to a tractor. The latter had big iron wheels, with spade lugs that dug into the ground to give drive and impetus. To travel on the road they had to be 'banded up'. This meant bolting on two half-circles of flat iron to each wheel to cover the spade lugs and give a smooth surface on which to travel. Mowers, that could be operated by the tractor driver on his own, did not appear in any quantity until the early thirties.

At the time of which I write, except for one-year and very occasionally two-year leys, all the hay was produced from permanent pastures, rich with a variety of clovers, vetch, a wealth of wildflowers and a large assortment of grasses, many now totally discarded from the specialised grass mixtures of the present day. Sprays were unheard of, resulting in an abundance of wildlife. It was not unusual to see ten or a dozen different varieties of butterflies and moths rise from the swathe as it fell. Field mice, voles and shrews were to be found in profusion, normally avoiding death by flattening themselves below the level of the cutter-bar, or scuttling down their holes. A host of bees and other insects would be disturbed by the mowers, as they bumped and rattled on their way. These attracted numerous birds, particularly swallows and the like, that would feed off the rich harvest that rose from the sweet-smelling hay. Care was always taken by the carters, as they rode on the seat at the back of the mower, to keep a careful watch for pheasant and partridge nests. If seen in time they were always mown round, and whichever gamekeeper was involved would be informed as soon as possible. Normally the beat keeper would visit the field several times while it was being cut.

There are few more tranquil sights than two or three horse-drawn mowers working in a field, on a hot June day. The sweet aroma of fresh cut hay, the chatter of the knives and the chirp of grasshoppers – of such things poets wrote, but it was not always peace and tranquillity. I

remember standing on the brow of the big field, opposite Lynch Hill Farm, watching a heavy crop of clover
being cut. There were two mowers at work, one being
drawn by a pair of Suffolk Punches, the other by the last
remaining Shires on the estate, a lovely matching pair of
greys standing 17½ hands high and named Hector and
Hercules. Charlie Coxhead always worked them. As I
stood quietly absorbing the beauty of the countryside and
subconsciously watching the greys as they plodded along
the far side of the uncut hay, I suddenly realised all was
not well. Seconds later I saw both horses rear and faintly
heard Charlie's cries of 'Whoa, whoa', but it was to no
avail. They turned and were off at a gallop, straight
through the middle of the crop, Charlie hanging on for
dear life! I realised I was standing directly in their path –
they were heading for their stable – but it was obvious
that there was nothing that I could do to stop their mad
flight, so I hastily moved. Just as they thundered past,
one wheel of the mower hit a bump and Charlie was
catapulted into the air. Thankfully he fell backwards. Had
he been thrown forwards nothing could have saved him.

He picked himself up immediately and started to run
after his beloved horses. I joined him. As we breasted the
rise in the field, the runaways reached the gate leading out
onto the lane. They turned too sharply and, with a sicken-
ing crash, the mower hit the gatepost. Both horses fell in a
heap. The draw-pole broke, with a crack like a rifle shot,
as the greys scrambled to their feet and continued their
flight for the stables. As we arrived they were standing by
the door, dripping sweat, eyes staring, flanks heaving.
Blood trickled from a nasty gash in Hector's side, from
which protruded a large splinter of wood. It was obvious
that he was going to need attention from the vet.

'What happened?' I asked Charlie.

Fighting for breath he gasped, 'Wasp nest.'

I looked again at the horses and could see a number of lumps starting to appear on their hindquarters and sides, from the stings they'd received. I helped Charlie unravel the chains and harness and get the horses into the stable, before rushing off on my bike to the estate office to ask the secretary to phone the vet. On the way I met father and stopped briefly to tell him what had happened. From the office I went to Britwell, found Old Jay and told him that father had said he was to go at once to Lynch Hill field to exterminate a wasp nest.

When I got back to the stables father was talking to a very shaken Charlie Coxhead. We walked out to look at the mower. It was not a write-off, but it would have to be completely dismantled and Old Jay's forge would be kept going for quite a few hours as he heated and straightened numerous sections. It looked as though a number of new parts would be required as well. Charlie Davis would be busy, too, making a new draw-pole.

Just then the vet, Aubrey Ward, drove up. Fortunately, he had been at home when the phone call was put through. Hector was brought out into the yard and the vet examined the wound. There was about two inches of the splinter, an inch wide, sticking out of the horse's side. Mr Ward stood looking at it, pursing his lips, before turning to father.

'Nasty, very nasty indeed, but it could have been worse. It's gone up the outside of the ribcage. Had it gone inward goodness knows what damage it could have done. That's driven in hard. It's my bet that'll take some shifting.' He touched the end of the splinter. Hector snorted and backed away. The vet went to his car and came back with a pair of shoeing pincers.

'I'll want these to get a good grip on it. Now we'll need help.'

Father went off to the cowshed and came back accompanied by Jim Light and one of the under-herdsmen.

'Good, now put a twitch on him, screw it up really tight. One of you hold up his foreleg and two of you hang onto his head. The poor old lad's not going to like this.'

He didn't and Mr Ward had several goes at extracting the splinter before it finally came out – it was 17½ inches long! I kept it in my 'museum' for a number of years. Fortunately it did not seem to have severed an artery and there was little bleeding. A few stitches were needed, but the wound was left slightly open to drain. There were no antibiotics in those days. From then on it was a case of keeping the wound clean and letting nature take its course. Hector had just been returned to the stable when Old Jay arrived. He'd treated the nest and, swearing, told us what a nasty lot they were – he'd been stung twice. No one disagreed with him, least of all Charlie.

Aubrey Ward said he had to be on his way, to go and see a goldfish! Apparently, the elderly lady who owned it had called him at 3.30 a.m. She had not said what was wrong, but insisted that he come at once. Grudgingly he had gone. When he arrived the lady took him into her drawing room and there, floating on its side, in a large glass bowl, was a goldfish. Its name, he was told, was Moby. It appeared that, for a man known to be somewhat volatile, he behaved with quite amazing decorum. He asked for the kitchen, took the bowl and poured two-thirds of the water down the sink. Then he slowly ran in cold water, aerating it well and thus improving the unfortunate Moby's supply of oxygen. He returned to the drawing room, made a great show of looking through his bag for the right pills, finally taking out an aspirin and dropping it into the bowl. He then asked for 2 guineas. His normal fee, for a night call, was 7s 6d! That, he thought, would be the end of the matter, but at lunchtime he had received a phone call to say that Moby had quite recovered, but that his owner would like his 'doctor' to call and just check that all was well. She also said that Moby had a little thank you present for him. We subsequently learned that the present turned out to be a half-case of whisky and led to our esteemed vet saying he wished he had a few more clients with goldfish, for they were far more profitable than pigs!

After a hay crop was cut it would be turned several times to dry it. I must have spent hundreds of hours as a boy, riding peacefully on the swathe-turner, kicking over the rows. It was a most relaxing pastime and one that allowed much time for meditation. Once dry the hard work began. First the hay would be cocked, that is, forked into heaps about 5–6 feet across the base and 4–5 feet high. Once things had reached this stage the crop was at least partially safe, because it was shower proof. As soon as possible, after it had been cocked, it was carted and ricked. This entailed pitching it, with two-grained forks, onto the big four-wheeled wagons to be carried to the rickyard. I learned, when I entered my teens, that this was a back-breaking, hand-blistering job. But at the end of the day, as one looked at a cleared field, it could be most rewarding knowing that one more rick was completed and winter fodder safe for livestock.

As a boy I loved it when the loaded wagons had to go up a steep hill and a trace horse was required to give that extra power to take it to the top. Whenever possible, I volunteered to ride the horse back to the field, to be ready to help with the next load. Once when hay was being drawn from Grey's Two up the hill past Burnham Grove, I was doing just this. I was about nine or ten at the time, still wearing shorts as all boys did in those days. As we left the road and entered Grey's One my thoughts were on circuses. I decided to stand up on old Tom's back, a placid horse who was coming to the end of his working days. To begin with I was rather tentative, holding onto the hames attached to the collar. I became braver and, as Tom slowly plodded across the field, I let go and stood up. It was easy! I jigged about a bit, even turned round, then suddenly I slipped, but I never made it to the ground. A hook on the end of the trace chains, hanging from the hames, caught in the leg of my shorts. There was a loud tearing noise before the hook remained firm somewhere around the crotch. There I was, literally suspended by the seat of my pants, dangling some three feet above the ground!

Tom started to graze, while I remained helpless, gently swinging to and fro. I called out, but it was no good – I was too far from the field gang. It must have been about ten minutes before I eventually heard a wagon coming. I was just able to peer round and saw it was George Wright, who wore strong pebble glasses and normally worked as carter for the flock, but at hay time it was a case of 'all hands to the pump'. It would not be unfair to say that George was not the world's brightest, but he was a steady worker and had a great loyalty to the estate. I called out, but it was at least a minute before he saw me.

'What youm doin' thar bhoy?' I explained that I had slipped and would he please lift me down. He stood thoughtfully eyeing me for a minute and then enquired, 'Youm goin' ter field?' I replied that I was and would he please get me down. He did not seem to hear what I said, but instead said 'Ah, then youm a'right.' With that he attached Tom to the back of the wagon, climbed back on and drove off for the hay field, some two to three hundred yards away, leaving me swinging gently by Tom's side, still firmly hooked by the seat of my shorts!

When we reached the field it was Jim Brookling, Tom's younger brother, who saw my predicament and came to my aid.

'What youm at Garge? You could o' hurt the lad.'

'Naw, him'd plenty o' air aroun' 'im.' My feelings were never very warm towards George after that! One good thing came out of that incident, though. I was tall for my age and father decided that the time had come for me to graduate to long trousers. Next day I accompanied him and mother to Maidenhead, where we went to a very new modern store that had opened at the beginning of the week and where nothing cost more than 5s (25p). It was called Marks & Spencer. Two pairs of grey flannel slacks were bought for me, costing 3s 6d each.

When the wagons reached the rickyard they would be drawn up alongside the elevator that carried the hay up onto the rick. This was driven by a petrol engine, frequently half-submerged beneath hay that had fallen off the conveyor. It only needed the engine to backfire, just one spark, and there would have been a roaring inferno within seconds. Every year, up and down the country, there were fires started in this way. The Burnham estate was no exception, for in the very dry summer of 1928 just such a fire started at Britwell. The rick was being topped off and the three men doing this just made it to the ground before the ladder burned through. Thanks to the speed and efficiency of the Burnham fire brigade it did not spread to any buildings except the timber store where, much to Charlie Davis's fury, a large quantity of well-seasoned ash and oak was totally destroyed. The control of the fire was largely due to a number of volunteers who accompanied father back onto the rick, once it was well soaked, and threw off loads of smouldering hay, which was quickly hosed down and rendered harmless. Fortunately such happenings were rare. The normal cause, for a hayrick to go up in flames, was internal combustion, caused by the hay being carted too soon and the rick overheating. I am glad to say that this never happened at Burnham, nor on any of the estates that I ran later in life.

The rick builder had a most responsible job. He would have two, sometimes three helpers according to the size of the rick, pitching hay to him as he built it, bringing the sides up absolutely vertically until starting to draw in for the roof. It was a very skilled job. If done incorrectly the rick would tilt and could, at the worst, slide over into a heap or, at best, require 'policemen' – great baulks of timber to prop it up. When this happened it caused a considerable loss of dignity to the builder! But I can only remember 'policemen' being required a couple of times at Burnham.

There were several excellent rick builders employed on the estate, 'Gunner' Stingmore for one. The sides of his ricks were so straight that anyone could have been excused for thinking he had used a plumb bob. A feature of rick building that always fascinated me was how someone like 'Gunner' could take a look at a field covered with cocks of hay and, from what appeared to be a cursory glance, know how big to make the base of the rick. Ninety-nine times out of a hundred he'd be right, almost to the last forkful to top the rick off. Once the latter had settled, the sides would be raked and pulled. The hay that came from this would be put along the ridge and the rick thatched with good wheat or rye straw and, if the hay had been made correctly, it was safe for two, possibly three years. Towards the end of the twenties hay loaders, that were attached to the rear of the wagons, came into vogue. They lifted the hay straight from the rows and deposited it in the wagon. So, after generations, the traditional way of making hay began to change.

At Burnham, the majority of the soil was light and gravelly, helping to bring about the early ripening of the crops. Harvest followed quickly on the heels of haymaking. One of my earliest recollections of visiting a corn field, at harvest time, was seeing three men 'opening up' the field, that is, cutting a road round the edge to let in the binder, as the reaper and binder was referred to in Bucks. This was sometimes done with a scythe, but on the estate those involved seemed to favour a hook and hank. The former was a sickle, the latter a hook of wood, usually hazel, used to draw the corn towards the reaper as he cut it some 5–6 inches from the ground. Two men would cut, while the third would gather the corn up into sheaves and deftly tie them with lengths of straw, leaving the heads of grain clear and undamaged. Every ear of corn was important. There were no grain mountains then, it was a precious commodity! The field would be 'opened up' the day before the binders came in. These normally had a cut of 4 feet 6 inches, so it took quite a while to cut a 20 acre field. It was a hard slog for both men and horses. What time they started was determined by how quickly the dew was gone. The crop had to be dry, for if the canvases which carried the cut corn up into the machine to be tied into sheaves were wet, they would slip on the wooden rollers and the machine would cease to function.

Once I started shooting, harvest became an exciting time. Rabbits, as I have already intimated, were a real scourge and the fields of corn would be full of them. Part of the gamekeepers' work was to be present and shoot them as they made a run for their burrows in the hedgerows or adjoining woods. Great care had to be taken when doing this, for there could be several gangs of men stooking, that is setting up the sheaves like the roof of a house: grain at the top, the butts of the sheaves on the ground. According to the crop and the district, so the number of sheaves to a stook varied. At Burnham it was eight of wheat, ten of barley and six of oats. The latter was usually cut while the straw was still slightly green, so it was necessary to allow plenty of time for the grain to harden and the straw to dry. The latter, unlike wheat and barley straw, was useful fodder. There was an old and true saying that oats, once cut, should hear the church bells ring three times before being carted and ricked. This old adage contained much truth for oats, if carted too soon, would heat in the stack and not infrequently catch fire.

Shooting in the harvest field required the highest standard of safety. Not only were there the gangs of stookers to watch out for – and there could be four or five of these, particularly in the evenings, when everyone was anxious to put in some overtime – but there would also be two or three binders moving round and round the crop. Somewhere in the field would be Tom Bunce, known as Tom Peg-leg, sharpening knives for the binders, if he wasn't stooking. A gun stood at each corner. It was some years before I was allowed to take a corner on my own. Prior to that I used to stand with either Bob or George. We used to keep tight up to the uncut corn, shooting away from it as the rabbits made a break for home. I became very accurate with my .410.

The rabbits were very much a part of the harvest, for they sold from 1s to 1s 6d each. Over a normal year something in excess of 3,000 rabbits would be sold off the estate, and one year the number exceeded 4,000. The most I ever remember being shot out of a piece of corn was 417 from 12 acres of wheat. It is not hard to visualise the damage such a horde could do. With wages around 38s a week, the receipts from rabbits would pay the year's wages for four or five employees. During the harvest all the staff received a brace of rabbits, and again during the winter. At 3s a time this doesn't sound as though it were of any great worth, but I can assure you that it was. To appreciate fully the value it should be said that at Christmas each employee received a voucher for 6s to obtain meat from Johnny Hall, one of the butchers in

Burnham. For that money the recipient could obtain a 6 pound joint of topside or rolled sirloin and have a real blow-out over the holiday or, as some did, buy lesser joints and make their gift last for several weeks. A family like the Goodchilds, at Biddles, where the father and five sons worked on the estate, would spread their gifts over several months.

Because of their value, it was most important to keep a careful tally when shooting rabbits in the corn fields, and make sure all were collected at the finish. The stubble was sufficiently long that it was not always easy to see them and so it was vital to know how many one was looking for. I had graduated to a twenty bore and was allowed to take a corner on my own, when suddenly we seemed to be regularly two or three rabbits short in our combined count. This was unusual to say the least. To begin with we put it down to a miscount. However, after it had happened five or six times and we were being particularly careful to note every one shot, Bill Yeoman mentioned the matter to father. Bob, who was undoubtedly Bill's mental superior, noticed we were never short when a piece of corn was finished early in the day. Further, it seemed that rabbits were never missing unless the corn being cut was adjacent to a wood, or cover of some sort. Being wise in the ways of poachers and scalawags in general, Bob kept his eyes open. The only discovery he made was that Tom Peg-leg seemed to get 'the call of nature' more frequently than others and would depart several times in the course of an evening to a neighbouring wood. Bob also noticed that it was only when Tom was stooking, which he liked to do on his own, that we were short and that when the field was merely surrounded by a hedge Peg-leg's bladder seemed as normal as his companions'!

One evening, not long before finishing a piece of wheat, adjacent to The Belt in Big Field, father arrived with his gun; so Bob went off about his work – at least that's what everyone thought. However, having reached the lane, Bob doubled back and made his way up the centre of The Belt and, unseen, positioned himself so he could observe what was going on. The men were well up behind the binders with the stooking, the main flush of rabbits broke from their dwindling cover, and the majority were shot. Minutes after the final sheaf left the machine, the last stook was completed. As we gathered up the rabbits, the men said good night and headed for their bicycles, which included Tom's, on the far side of the field. (The latter had a rest for his peg-leg and, with the aid of a fixed wheel, could drive his bike along as fast as anyone.) As Bill, George and I hocked and checked the number of rabbits, I saw Tom again making his way to The Belt, for at least the third or fourth time that evening! We finally decided that we had definitely collected everything and yet we were three rabbits short. We checked again, but the answer remained the same.

We were just about to go when Tom reappeared, followed almost immediately by Bob. As the former reached us he touched his cap to father.

'Night, sir. Reckon 't will be a rare ol' day termorrer. Me ol' wooden leg's not aching at all.' This was Tom's stock joke.

As he moved off Bob caught up. 'Hang about, Tom' Then, turning to father, he continued: 'I think you should tell Bunce to drop his trousers, sir.'

'Do what?' Father, to say the least, looked surprised.

'Drop his trousers.'

'That's what I thought you said.'

Tom joined in: 'Now look youm here Bob Hedges, what youm on about? What kind o' a chap do youm reckon I am?'

'A thieving rogue Tom Bunce, that's what you are. Now

are you going to drop 'em, or do you want I to do it for you?' Bob was a very powerful man and his prowess in dealing with truculent poachers was legendary. Cursing and nearly in tears, Tom undid his braces and belt, his trousers falling to the ground. The mystery of the missing rabbits was solved, for there, hanging from nails driven into the wooden leg, were three rabbits!

Tom was sacked on the spot. He didn't consider this an injustice. After all, if it had happened only a few decades earlier, he might well have been deported, for his offence would have been deemed worse than poaching. He had been caught red handed stealing from his employer. In fact, he was very grateful that the police were not called. Jobs, however, for a one-legged farm worker were few and far between. Three weeks later father, having discussed the matter with Edward Clifton-Brown, who was highly amused by the whole incident, sent for Tom and reinstated him. Tom worked on the estate until retirement through ill health, but at harvest time he was never allowed to forget his misdeed. It was not unusual to hear one stooker say to another within Tom's hearing: 'Do youm reckon thar'll be any ol' rabberts goin' up ter roost ter nite?' Or 'Youm reckon ol' Tom'll 'ave ter shave 'is 'ol peg ter nite?' Tom would only smile. He never took offence and was grateful for his job.

Once carting started, in a good summer, it stretched the labour force to its limits. Frequently there would be two gangs carrying and ricking corn, while more was being cut and had to be stooked. Often it would be dark by the time the binders stopped and the last sheaf was in the stook, the latter job being done at these pressure times by the stockmen, many of whom had been at work at 5.00 a.m. A long hard day, but it was only for a few weeks and pride and the much-appreciated overtime were the spur. The corn ricks had to be built with even more precision than hay, the heads of the sheaves being inwards and sloping upwards, so that the wet could not seep in. The dry sheaves could easily slip and they had to be truly dry – well, barley and oats did, wheat could be damp and come to no harm. In fact I have seen the water running out of the bottom of a wagon bringing in the last load of wheat to top a rick. Once the ricks had settled, Jim Stannett, a self-employed thatcher, would arrive in his pony and trap. Quickly, with the aid of two men to wait on him, he made the ricks safe against all weather. When the last thatching peg was driven home there was a sense of both relief and achievement. From then on the only damage, until the time came for opening up for thrashing, would be from rats and mice.

While the main harvest was going on there was a smaller one taking place as well, equally important to those concerned. A number of the staff living on the estate had a few hens. As soon as a field was cleared, many of the wives would be out on the stubble, gleaning – that is, collecting the heads of corn not gathered in by the binder. Wheat was the most popular and supplied a useful adjunct to the feeding of hens. Nothing in those days was wasted.

8

Holidays

AS soon as the harvest was safely in we went on holiday.
My grandfather was rector of Beaford, a small and
very rural village in North Devon. Each year we stayed at
the rectory. It was a large rambling house, with extensive
grounds which included a very fine walled garden. In late
August this was a veritable paradise for small boys with its
abundance of peaches, apricots, green figs, big golden
plums, juicy pears and a variety of apples. The greenhouse,
too, was stocked with goodies, including both black and
white grapes. The house was approached by a long winding drive, partially wooded on one side. About halfway
along it was a saw pit. Here, in years gone by, great trees
were cut into planks with crosscut saws, one operator in the
pit being continually showered with sawdust, the other
on, or straddling, the log that was being sawn. I can
just remember the pit being used, but only for cutting
up the tons of logs which were required to bring any
semblance of warmth to the rectory in winter time. There
was no electricity. Every night when going up to bed
we would take a candle off the table at the foot of the
large and imposing staircase. From the front of the house
was a long passage leading through to the kitchen and
numerous storerooms, beyond which was the back kitchen,
big enough for Ralph and I to have cycle races round it on a
wet day when we could not go out. Over this and the dairy
was a private chapel – there was no shortage of space!

My grandfather was a great personality. His parish, I recall, was about 6 miles by 4. He did all his visiting on foot and this was done on a very regular basis. All his parishioners knew him. Sunday mornings, when he went up into his pulpit, he would slowly and deliberately survey the congregation; absentees were noted and on Monday morning, rain or fine, he would set off to enquire as to the reason for their not 'being on parade'. Every second Wednesday he would walk the 4½ miles to Torrington station and take the train into Bideford, a picturesque journey along the Torridge valley. The little engine pulling the train was aptly named Rose O'Torridge. Having reached his destination he would have a haircut, change his library books at Boots, do whatever shopping he wished to, lunch at Bromley's Cafe and set off home. On a fine evening, his return was heralded by his rich baritone extolling the virtues of the Lord, as he sung his chosen hymn with feeling and exuberance as he walked down the drive. He had a fine voice and enjoyed using it. He was an excellent cleric, although he always described himself as an 'unclerical cleric' – in fact he wrote a book which was published under the title of *Reminiscences of an Unclerical Cleric*. He was a brilliant fly fisherman and, thanks to his good offices, my father, brother and I could fish for trout on some 7–8 miles of private water on the Torridge. He was an avid chess player too, and a very fair judge of port. In the winter he and a retired general, who lived in Beaford, kept two games going, one at each house. They would meet several times a week, alternating between each other's homes, play until they had consumed a bottle of port and then go home. He thoroughly enjoyed his food and it was a family joke that when grandfather said grace it was all in one breath – 'Benedictus benedicat Ethel what's for dinner?' In my childhood I did not know that my grandmother's name was Ethel and thought the latter must be God's housekeeper!

My grandmother was a quite amazing lady – tall, slim, snow white hair and the most vivid and penetrating blue eyes I have ever encountered. To say that she was a character would be to minimise her personality. She was very actively involved in parish affairs but not, as one might imagine, with the W.I. or church flower arrangements, dear me no, she left those sort of namby-pamby pastimes to other ladies within the community. She ran the Boy Scouts, the Girl Guides and was secretary of the local football club, frequently going to matches and never slow to give strident advice and encouragement from the side line. Annually, for many years, she went to camp with both her scouts and guides and at seventy-two years of age spent a week under canvas at some major scout and guide jamboree, at which representatives from her Beaford troop won the top award for tracking. Granny had coached them through the abundant woodland on either side of the Torridge!

It was an unwritten law of the parish that all boys joined the scouts – they could volunteer or be pressganged! Off the passage leading from the front of the rectory to the kitchen were a number of rooms, one of which was given over to the scouts. One afternoon I was making my way along this when I met my grandmother, coming from the direction of the kitchen, grasping a very recalcitrant lad of some 8 or 9 years by the collar. She was wielding a signalling flag. 'Your mother says you are to join the cubs.' Whack! The handle of the semaphore flag was brought smartly down across his buttocks. 'I say you will join the cubs.' Whack! 'You are going to become a good scout and enjoy it.' Whack! 'Aren't you?' Whack! 'Yes ma'am, yes I promise.' The flag was lowered. He did become a good scout, enjoyed it and eventually became an Assistant Scout Master. He also grew up to be a man of considerable talent, was devoted to my grandmother and did much to help her in her old age. There might well be a lot to be said for old fashioned methods of recruitment!

My early visits to Beaford were made in a bull-nosed Morris, the first time in a two-seater with a dickey. For those who are not acquainted with this particular part of an automobile, imagine something akin to a car boot, hinged from the bottom, that opens up to provide a seat. Mother, who did not drive, always rode in the dickey. If it was fine either Ralph or I rode with her, if wet we were under cover in the front with father. It was 183 miles from Burnham to Beaford, these days no great distance, but in the early twenties it was a day's drive. We would leave at 7.00 a.m. and reach Yeovil for lunch. Of course there were no by-passes or ring roads around towns and the word motorway was unknown. We would arrive at Exeter in time for tea. This for Ralph and I was the high spot of our journey to the West Country; we would fortify ourselves, at the Rougement Hotel, with a real Devonshire tea before tackling the last leg of the journey into what was truly rural England.

There were only two petrol pumps between Exeter and Torrington and only two between Bideford and Bude in North Cornwall. We, and most other motorists, carried two spare cans of petrol, which used to be held in a frame attached to the running-board. Having topped up with petrol costing, if I remember rightly, 10d per gallon (slightly less than 5p in today's currency), we set off on the final stage of our journey. I well remember the first trip. It was a glorious August evening and I was riding in the dickey with mother, wide-eyed, excited and, indeed, even more excited when we ran over a duck in the middle of the quiet little town, or was it village then, of Crediton! Brakes in those days weren't what they are today. We arrived at the rectory at 6.30 p.m. after a long and tiring day.

Father's home had been at Minehead and, in his youth, he'd spent many happy hours roaming Exmoor; he was a naturalist and fisherman of no small ability. One year, the morning after our arrival, he took me off to explore the stream that ran along the combe below the rectory. For reasons I have long since forgotten Ralph did not come with us. It was a warm damp morning as we made our way down across the meadows. The sun was trying hard to break through and when it did it was going to be hot. A pair of buzzards was wheeling and mewling overhead as they rode the thermals, a raven soared above them. It was truly perfect peace, as it had been for many centuries. By the time we reached the stream the sun was shining, highlighting the profusion of wildflowers that covered either bank, as butterflies flitted from flower to flower. In a matter of minutes I saw several varieties of Blues, Painted Ladies, Commas and a host of Peacocks and Tortoiseshells. As we stood there a dipper landed on a nearby rock, immediately breaking into its sweet warbling song before suddenly submerging itself in search of food.

Father surveyed the scene. It was obvious, even to my untutored eye, that it would be impossible to cast a fly for the brown trout that inhabited the stream in large numbers. He turned to me: 'Take off your shoes and socks.' He did the same, put our things into a knapsack he'd slung over his shoulder before leaving the house and rolled up the legs of his trousers. There was no need for me to do that, for I was wearing shorts. 'Now I'm going to teach you to tickle trout.'

'What, like Jan Ridd in *Lorna Doone*? (I had recently read R. D. Blackmore's famous story and had lived every minute of it.)

'Yes, just like Jan.'

Tickling trout is, of course, strictly speaking poaching but, in those days, with every little stream in that part of the country full of brown trout, virtually all unfishable with rod and line, being so overgrown, no one looked upon it as a serious offence. Trout always lie facing upstream. Tickling them is an art and it is literally what one does. Slowly and quietly we made our way up the brook, as father gently felt under the overhang of large stones. I followed, tense with excitement, as the cold water rippled around the calves of my legs. I saw father smile. I felt sure he'd found a trout. Almost imperceptibly his arm moved forward. I could visualise his fingers gently caressing the underside of the fish, as they eased ever nearer the gills. Suddenly, with a swift movement, his hand appeared clasping a lovely half-pound trout, which he quickly slipped into the bag hung around his neck.

Then it was my turn. Father found a trout and, with the minimum of disturbance, withdrew his hand and whispered to me to try. Alas, I was far too impetuous and, in a flash, my quarry had gone. However, having learned what to do, I started to work one side while father continued on the other. Several more went into his bag before I found one – and when I did I realised that it was big. Hardly daring to breathe I moved my hand forward, slowly, oh so slowly, just as I had been watching father do. Now! I grasped the slippery fish and whipped it out of the water. It wriggled frantically and seemed certain to get away. I made a frenzied grab with my other hand and, as it closed around my quarry, I sat down unceremoniously in the water! It still seemed as though it might escape me and, having no bag, without further ado I dropped it down inside my open-necked shirt and made for the bank. Father followed, tears of laughter running down his cheeks. He produced a small spring balance and weighed my first trout. It was three-quarters of a pound – truly a monster for such waters!

I was soaked up to my armpits, but the sun was very

hot and father decided I would come to no harm. I stripped off and he wrung the water out of my clothes. They soon dried. I 'tickled' several more trout, but was not allowed to keep them, because they were too small. We continued until we had twelve, well father had eleven – I only had one, but it was the biggest! It was only 11.00 a.m. when we returned to the rectory and entered through the back door. Olive, the cook/general, was in the kitchen. She produced a large dish on which to lay out our catch – I thought they looked magnificent. Having washed our fishy hands, we made our way through to the drawing room, where the rest of the household were gathered. I carried in the fruits of our labours.

My grandfather meticulously inspected our catch. Then, beaming upon me, he placed a hand on my head and said, 'Verily, verily my child you have the makings of a most prodigious poacher!' Turning to my grandmother he added, 'Ethel, we'll have these for dinner.' At least, I thought, he won't have to add 'Ethel, what's for dinner' to grace today. But he did. Old habits die hard.

It was later that day that we heard of Peppercombe, for the first time. It was a stretch of beach that lay between Westward Ho and Bucks Mills and, according to grandfather, was the home of massive prawns. His informant was Mr Mudd, a strange, shy little man who lived with his sister within the confines of the parish. Apparently Mr Mudd had a friend with a car, whose parents lived at Bude. When the friend was visiting Bude he would drop off Mr Mudd, who would walk the mile down to the bottom of the combe and then spend the day on the beach, prawning. The only trouble was, grandfather informed us, that the 'road' down to the bottom of the combe was impassable for motors.

Next day father decided to investigate. We parked by the pub at the top of the road, close to a large sign saying

'Impassable for Motors'. It was more than an hour before father returned, hot, but looking triumphant. He indicated the sign – his only comment was 'Nonsense!' – then climbed back into the car and away we went. Mother, a non-driver, as I have already said, protested vigorously. I must admit that, in places, it was a bit 'hairy' and the gradient was certainly down to 1 in 4, possibly even less. Father, however, being, as he liked to say, 'an Exmoor man' had a supreme contempt for West Country hills, and we reached the bottom safely. (It was said that he was the first person to ride a motor bike up the notorious Porlock Hill, around, I believe, 1909 or 1910. I gather he achieved this feat on a belt-drive Douglas.)

There was a large area of grass at the bottom where the car could be turned with ease and parked in the shade. We boys tumbled out to explore and were greeted within minutes by the farmer, whose homestead was in splendid isolation at the bottom of the combe. He seemed very old, had a grey beard, a weatherbeaten wrinkled face and a broad and friendly smile. He gave us two large rosy apples and immediately became our friend for life. In the years to come we never visited Peppercombe without receiving an apple each. The farmer chatted to mother and father for a while in the broadest West Country dialect, then we headed for the beach. It was only a short walk past a wooden bungalow, used occasionally at week-ends by a lady from Bideford. It was nearly low water, exposing patches of sand and pools, as well as rock and pebble, the latter being similar to that which forms the Pebble Ridge at Westward Ho. As far as the eye could see towards Clovelly, and back towards the estuary of the Taw and Torridge rivers, there was only one person in sight – and this was in the third week of August! The lone figure was following the tide back, obviously working the pools for prawns. Father, being well versed in such arts from his boyhood days at Minehead, departed to seek local knowledge. It transpired that the lone figure was a man called Bagotty from Bucks Mills, a professional fisherman who not only went prawning and worked shore lines, but also crewed in one of the herring boats out of Clovelly during the season. The outcome of father's chat proved most rewarding and, in the nicest possible way, it was suggested that we should 'fish' further along the coast towards Westward Ho, below Portledge House – this was long before it became a luxury hotel.

After several hours exploring we made our way back up to the car and again had a chat with our farmer friend. From him we learned that there was a path along the cliff top that would take us right to the beach at Portledge. He even took us round and showed us where it started, at the back of his house. We then set off up the combe, mother exclaiming volubly at father's folly in bringing the car down the stone track and, at the same time, telling him what he was doing wrong. She undoubtedly had a Certificate in Advanced Back Seat Driving! Father, fortunately, was totally impervious to this distraction, concentrating on the task in hand and observing the temperature gauge fitted on the radiator cap. When we reached the top, it had to be admitted, we had a good head of steam up! We parked by the pub, while father went in search of a can of water. When the poor old car or, to be truthful, new car – it was a four-seater bull-nosed Morris Cowley – had cooled off, we headed back to Beaford Rectory. So started many years of memorable days spent on the beaches at Peppercombe and Portledge. First it was the Morris that passed the impassable, then, in later years, an Austin 12 which was followed by an Austin 16.

Two days after we first ventured down the combe we were back. Three of us were armed with prawn nets, while Ralph had his butterfly net and a killing bottle. Father had an additional piece of equipment – a large fish hook, with the barb filed off, lashed to a stout five foot garden bamboo. He would not tell me what it was for and, aggravatingly, kept saying 'wait and see'. As we made our way along the cliff path, laden with picnic baskets, mother's deck chair and sundry other items, the sun was blazing down. The sea shimmered in the heat haze and, high above, we could hear the cronk of a pair of ravens as they glided majestically towards Hartland Point. It was obvious that Ralph was not going to be short of interest. The air seemed full of butterflies, as they rose from the wealth of wildflowers abounding by the path. Several varieties of Fritillaries mingled with Red Admirals and numerous other species. In one place there must have been several hundred of the beautiful Cinnabar moths, on a large patch of ragwort, while a profusion of small moths arose from any patch of long grass that we walked through.

We reached the beach and 'made camp', agreeing that we should meet for our picnic lunch between 1.30 and 2.00 p.m. Low water would be around 12.30 and father was most anxious to follow the tide right back, for it was THE big spring tide of the year. Then, having seen mother happily started in a large pool – she enjoyed paddling around in pools close to the high water mark – father and I made our way out towards the sea, while Ralph headed back to the fields. This was my first attempt at serious prawning. I watched father carefully, as he worked his net around under the seaweed and ledges of rock. When he lifted it out of the water there must have been twenty or more prawns in it, together with a number of small rock fish and green crabs. He picked out seven or eight whopping great prawns and put them in a bag hung round his neck. Then, to my horror, he tipped the remainder back into the water! They were quite as big as some I had seen in Macfisheries!! When I asked why he had done this, he told me that one should always look to the future and preserve stocks of wildlife, whatever their form.

I started working my net and every time I lifted it I had two or three keepable prawns. I was fascinated not only by my catch, but by the immense variety of rock fish and small crabs. We worked steadily, our bags becoming ever fuller; then father took out his watch and announced that it was half an hour to low water. Suddenly he called me over to him and told me to hold my net below a hole in the rock. Carefully he inserted the hook on the end of the bamboo into the hole. I was watching him carefully, his face a study of concentration, and he smiled.

'Ah, he's at home.'

I enquired what.

'Why, a lobster. Now when I draw him out, if I haven't got him properly, be ready to bring your net up quickly when I say.'

'Right, I won't miss.' I was almost shaking with excitement. Then, as father slowly withdrew the bamboo, I saw one claw grasping the cane, then the other – it was a monster, or so it seemed to me. Before I knew what had really happened Dad was holding a lovely lobster of some 3½ pounds. He looked at it, smiling.

'Not bad, not bad at all. Now we must find one to keep it company.'

We worked hard, following the tide back as pools became exposed. Some were all rock, others had sandy bottoms. I was wading knee deep across one of these when there was an unexpected movement under my foot and a swirl of sand. I shouted out in surprise.

'What's the matter?'

'I don't know. I stood on something quite big in the sand that's swum away.'

'Stand quite still. I'll come to you.' I must admit to

being a bit scared. I looked at the kelp swirling around in the sea, just beyond the end of the pool. It felt as though anything might live in that underwater jungle! Then father was beside me, and only seconds afterwards a lovely big plaice came and worked its way into the sand just in front of me. It wriggled until only its eyes and bare outline were visible. Father reached into his hip pocket and produced something that looked rather like a barbed pickle fork.

'Haven't used this since I left Minehead. Now don't move. I'm going to lean on your shoulder and reach round you.' This he did with the 'fork' in his hand, then he drove it forward. The next thing he was holding a plaice, that later proved to be a 1½ pounder.

'Now the question is, where's its mate? Incidentally, if you ever "start" a plaice in a pool, stand quite still, it will always come back to where it was lying.'

Father moved to the right, I went left, making as little movement as possible in the water. Then I saw it just in front of me.

'There!' I gasped, pointing at the sand about a yard ahead of me.

'Right, hold your net out.' Father dropped the 'fork' into it. 'Now it's up to you, move slowly, but, when you strike, put plenty of beef behind it.'

I grasped the fork, heart racing and prepared for the attack. I put so much beef behind it that I landed face down in the water and was totally submerged – but we had another plaice. I wrung out my shirt – I was only wearing it as protection against the sun – the rest of my gear was a bathing costume. This was before the days of trunks.

We continued prawning and it was nearly dead low water when father said, 'Enough. We've all we can possibly eat.'

100

'But, it's such fun. Can't we catch a few more?'

'No, I said enough. Don't be greedy, always think of the future. You go back to camp and take these fish with you. Here are some matches. Get a fire going, with some driftwood, like I showed you the other day. Then fill the kettle with sea water and have it boiling by the time I get back. I'm going to see if I can find another lobster.'

Reluctantly I did as I was bid. In those days children did not argue or, if they did, seldom twice. It had been a wonderful morning, to be repeated many times in the future. I speared numerous plaice over the years and, on one memorable day, a pair of soles. Years later, when I too was armed with a lobster hook, I started and captured a 14 pound conger eel. It is extraordinary, but when one disturbs a conger in a pool or even in the actual tidal water their first mad rush is towards the shore. That is your only chance. Miss it and you can say goodbye to your slippery quarry.

I arrived back at base. Mother was comfortably ensconced in her deck chair, reading a book – she was an avid reader. She greatly admired the plaice and my bag of prawns; then she showed me hers. I sniffed and then commented, 'Hmm, not bad, but you know you should leave those small ones for stock – always think of the future!' There's no doubt that all small boys can be odious at times!

When father returned his expression was enough to tell me that he'd found another lobster – he was positively beaming. At the same moment Ralph returned, his face alight with excitement. He told us that he had seen a pair of White Admirals and could have caught one, but decided against it as he already had one in his collection, for, as he truly said, they were quite rare.

'What do you want me to do with this kettle of boiling water?' I enquired of father.

'Why, cook some prawns, of course. You'll never taste better.' How right he was – I never have.

North Devon in the twenties could be described as almost devoid of traffic. It was quite normal to drive into Bideford and park, even in the High Street, outside the shop one wanted to visit. While in the country, particularly on by-roads, it was the exception rather than the rule to meet another vehicle more than once every two or three miles. If we did mother was almost certain to comment about the amount of traffic! Quite often we would spend a day on Dartmoor. Our route would take us through Winkleigh, a neighbouring village to Beaford. On hot sunny days, an ageing resident would have his armchair carried out into the centre of the street, so that he could enjoy the sunshine to the full. It was, I remember, a big old leather chair, with tufts of horsehair sticking out through holes in the arms and this led to me dubbing him 'Horsehair Harry'. As we approached his back would be towards us. When father gave a good toot on the car horn, 'Harry' would pull himself up onto his feet, wave his stick above his head and abuse father roundly for disturbing him. It was impossible to circumnavigate the chair but, invariably, passers-by would good-humouredly move it and, when we had negotiated the road block, replace it for the old gentleman to continue sunning himself!

Normally it was the second week in September when we returned to Burnham. As soon as the car was unloaded Ralph and I would spend several hours with our numerous pets, before departing to bed. I would be up in good time the next morning and head for the meadows at Lynch Hill, behind Tom Brookling's cottage, with a basket hung over my arm. It was the time of year for mushrooms. All the fields had mushrooms in them, but these particular two were always white with them and it was only a matter of minutes to pick all we could eat. I would be home in time for us to have them with our bacon and eggs for breakfast. They were delicious, their flavour vastly superior to the cultivated ones that are sold today.

9

Poachers and Poaching

IN the twenties poaching was no great problem. There were a few known miscreants who had to be watched, but they did it more for the sport than the money and were never a serious problem. There was one semi-professional in the village, who was reputed to have shot a gamekeeper near Windsor, but nothing had ever been proved. He, fortunately, for reasons best known to himself, liked to work further afield, although he could have been in the main coverts on the estate within 10–15 minutes of leaving his cottage.

However, this reasonably tranquil state of affairs was not to last. Industrial development was rapidly expanding on the southern boundary; in fact the estate farmed the land right up to the newly built factories. With these came an influx of labour from the depressed areas in South Wales and the North. A new and very large housing estate sprang up, only a few minutes' walk from Biddles Farm. Suddenly vandalism and poaching became part of everyday life; for the first time many of the locals were being forced to lock their doors when they went out of an evening, as petty pilfering became the rule rather than the exception.

A young and comparatively new member of the staff was getting married. There wasn't a cottage available on any of the farms, but he was delighted to find he could rent one on the new housing estate. He did not want to take any time off for a honeymoon – money was short – but he did ask for the Friday and Saturday morning, and if he could cut some turfs, on the estate, to lay a small lawn in front of his new home. Permission was readily granted and on the Friday evening two of Ike's work mates went over and helped him put down the grass. His was the only house in the cul-de-sac with a front lawn! Ike was married early the next day and he and his bride went to Cookham for their 36-hour honeymoon. When they returned to their home on Sunday evening the lawn, which Ike had proudly told his bride about, had gone – the turfs had been lifted and carted away!

Some of the poaching was of a very amateurish nature, more a nuisance than a real hazard to game. However, the poacher, into whatever category he falls, is a villain, a law breaker and not, as so often portrayed by some writers, a romantic character. Certainly there was nothing romantic about those who began to invade the estate in the thirties. Some were purely out for a bit of sport, as they put it when caught. Others, like Jim Sloane and Perkins (I forget his christian name) were professionals, well equipped and intent on taking all the game and rabbits that they could get away with. They were the ringleaders but there were a number of others, working with them, who were less persistent. To give some indication of the degree that poaching reached in the mid-thirties, seventeen cases from the estate were up before the magistrates at one sitting!

When in my teens, spotting and helping to catch poachers became an exhilarating and exciting pastime. There were some 2½–3 miles of private roads on the estate which, with the exception of one lane, were public footpaths. In addition there were several miles of highway crisscrossing the property, giving numerous points of access to both coverts and fields. Most of the trouble was on Bob Hedges's beat. He was assisted by George Devonshire and both men had to be constantly vigilant and on patrol. They must have worked quite phenomenal hours. The majority of the poaching consisted of ferreting, using purse nets, or long-netting for rabbits, although taking hares by coursing was popular too, particularly with Jim Sloane. He had a remarkable lurcher bitch, Rose, trained so all instructions meant the reverse of what he said; for example 'come here' meant 'get out', and so forth. She was very fast and could run a hare down with comparative ease. When she caught one she would give it one quick nip to kill it and carry it back to Jim. Hares were classified as game and so it was a greater crime poaching them than rabbits.

In the early days neither Jim Sloane nor Perkins turned their attention to pheasants, but the temptation was eventually too great. They were caught on a neighbouring shoot one moonlight night, snaring pheasants; that is, they had a rabbit snare mounted on the end of a long bamboo, with a cord attached. When an unsuspecting pheasant stuck its head out from the roost, to see what was happening, they slipped the noose over its head and pulled it tight. The penalty for night poaching was far more severe than that levied for a similar offence during daylight hours; both were lucky not to go to jail.

A haul of twenty or more rabbits could make the chance of capture an acceptable risk to the poacher. Probably, for every time he was caught, he would make several forays that would go undetected however vigilant the gamekeepers might be. Most of the time, when the keepers were successful, it became a boring routine of capture, call the police, charge the offender, appear in court and sit by while a ridiculously small fine was imposed. Frequently instructions were given by the magistrates to return both ferrets and nets, confiscated at the time of capture, to the offenders. However, after a while the bench hardened their attitude towards those who appeared before them with monotonous regularity and started to increase the fines, as well as appropriating all instruments used in the unlawful taking of conies or game.

One afternoon, in early May, I joined Bob as he was setting off around his beat. He had been in court all the morning, giving evidence. One of the offenders had been Perkins, with five charges against him. He was found guilty on all counts and stiff fines were levied. He was given one month in which to pay and the chairman assured him that should he not do so he would go to prison. We headed for Lamas Wood, where Bob had several tunnel traps. Gamekeepers wage constant war on vermin, all the year round, but intensify their activity in the spring. Having checked the traps, we continued along a hedge that eventually ran parallel to Hay Mill Pond.

Halfway along this a pair of carrion crows had nested in a high elm tree. They are one of the worst of predators and frequently very difficult to come to grips with; however, we had a plan. I was to walk out in the middle of the field to attract the attention of the bird on the nest, while Bob crept along the hedge. When I was opposite the elm tree I was to walk towards it, the assumption being that the bird would then fly out the other side. This is exactly what happened and Bob shot it. One down, one to go.

I secreted myself in the hedge, while my companion headed off to attend to more traps. We agreed to meet at the mill in half an hour. I had not long to wait before I heard the strident caw of a carrion crow. I crouched lower in the bush where I had hidden myself and peered up through the foliage. I could see the other member of the pair approaching. It circled several times, calling its mate, before finally coming in to the nest. I fired and, to my horror, missed. However, I was successful with my

second barrel and dropped the bird into the reeds surrounding the pond, in much the same area as the first one
had fallen. A good afternoon's work for, quite apart from
their attacks on game chicks, I had had 'a thing' about
carrion crows ever since I'd seen a new born lamb, very
much alive, that a crow had pecked the eyes out of,
minutes after it was born.

I joined Bob at Hay Mill, where he was chatting with
Rogers, the miller. Truly a mighty man was he, for he
would walk off with a 2¼ hundredweight sack of wheat as
though it only weighed as many ounces! We continued
round to the east side of the pond, where Bob wanted to
check a couple of partridge nests. In pre-war days it was
customary, on a well-keepered shoot, to find the partridge
nests, pick up the eggs as they were laid and put dummy
ones in their place. When the hen partridge had finished
laying and was happily trying to hatch out the wooden
counterfeits, the latter would be removed and the real
eggs returned. The reason for this performance was because it was not unusual for a partridge to be disturbed, to
a degree that she would desert the nest, or even at
times be killed. Should either of these eventualities take
place, then the eggs that had been picked up could be
redistributed around other nests.

We were walking along the top of a bank that ran
parallel to the pond and which was almost one continuous
rabbit warren, when suddenly Bob stopped, put his finger
to his lips and pointed down over the bank. There, lying
face down in the grass and nettles, were two men. It was a
hot afternoon. The man nearest to us had no shirt on and
was prostrate in a clump of nettles, but he never moved.
Purse nets covered a number of holes in the bank. Suddenly a rabbit shot out into one. The man furthest from us
started to rise and then sank back onto the ground. Bob
gave me a wink, moved on about ten yards, then started

to chat about a swan and her cygnets that we could see out on the water.

We'd been there about five minutes when the shirt-less man's shoulders started to twitch, then he began to wriggle frantically and suddenly, with a yell, he leapt to his feet. There, hanging from his chest, its teeth well embedded, was a large fitchy ferret. Blood was trickling down the man's well-stung chest – and he was none other than our friend Perkins! He leapt around, with consider-able vigour, before finally persuading the ferret to let go. Apparently he and his friend had heard us coming and dropped to the ground, hoping not to be seen. Unfor-tunately for him, apart from landing in a patch of nettles, he had quite unwittingly lain over a rabbit hole which just happened to be the one that the ferret chose when it decided to return above ground.

Perkins and his companion appeared before the next magistrates court; the former did not help his cause by telling the chairman that he was to blame, for had not such severe fines been imposed he, Perkins, would not have had to go and try and catch some rabbits to sell and pay the fines! His plea was not accepted and he went to prison. It was the last time we saw him.

Small and often trivial things can lead to the capture of the ungodly. I remember one evening in late October quietly cycling along the main road adjoining the north side of Big Field. I was just approaching the end of The Belt when I spotted two figures walking out across the stubble, about 80–100 yards apart. Although there was nearly a full moon, I could not see what they were doing, but I was certain they were up to no good. I sped off down the road. There was just a chance that Bob and George might be waiting in the shadows at the end of Crow Piece Lane, near a wicket gate that led into Cocksherd Wood. It was a favourite place of theirs when on night patrol. I was in luck and received an immediate answer to my muted call. I quickly explained what I had seen. Bob was most concerned and said it sounded as though the two men were using a drag net. This was something I had heard about, but never seen. It is a net 40–50 yards long by 12–15 yards wide, weighted at the back. A rope is threaded through the meshes at the front, extending either side to provide two 'tow ropes'. The net is opened out, then pulled across stubbles and open ground where partridges are known to have jugged down for the night. The two operators drag it along, keeping as much tension as possible to raise the front as high off the ground as they can. As soon as they hear or feel a partridge fluttering up they drop the draw lines and can frequently net a whole covey.

Plans were quickly made. George was to go back up the lane and head for the gap halfway along The Belt. Bob would go in where I'd seen the two men and I would go down Deep Lane, to cut them off if they should head back that way. It was George who saw them first but, alas, they saw him too and turned and ran towards Deep Lane. George gave a blast on his whistle – we all carried one – to let us know he'd sprung our quarry and set off in hot pursuit. He said later that he had been gaining on them when he caught his foot in a rut, measured his length on the ground and partially winded himself. Meanwhile, unseen by me, Bob was running across the field towards the opening that led out onto Deep Lane. I took up my position some fifty yards from this. I could hear running, someone fairly pounding along, but didn't know who.

Then two figures burst into the lane, turned and headed towards me. They were about twenty yards away when I switched on my torch, at the same time blowing my whistle. With a curse, they changed course and headed back towards Biddles. As I started after them, I heard someone rapidly approaching from across the field, so I waited. It was Bob. I had a carrier on the back of my bike, he jumped on and we set off down the lane, Bob giving a blast on his whistle to let George know which way we were heading. As we neared the junction with Biddles Lane we heard the wicket gate, into Gorse Field, slam. We were only seconds behind. Bob leapt off and opened the gate.

'Quick, give me your bike and I'll get round to the far side of the gorse.' (There was a big patch of this, after which the field was named.) 'Wait here for George and then the two of you quietly make your way down to the bottom corner. Listen for them. After that run they'll have gone to ground somewhere.' With that he was off.

A few minutes later George arrived, puffing and blowing. I told him the plan. When he'd got his breath back, we quietly made our way to our allotted place. We stood in close to the gorse to be in the shadow, for the moon made it nearly as light as day. There was just a hint of a breeze blowing towards us from the gorse – it would be easy both to see or hear anyone trying to make a run for it. We stood silently for a good 10–15 minutes, hardly daring to breathe.

Suddenly George leant towards me and whispered, 'Did you break wind?' (A certain propriety was maintained in those days; four-letter words were not yet the vogue.)

'No,' I replied, somewhat indignantly.

'Right!' To my horror George not only raised his voice, but at the same time switched on his torch and shone it on the gorse. There, only a few feet from where we'd been standing, protruding from the dense cover, were two pairs of boots! George stepped forward and gave a hearty kick on the sole of one of them. 'Come on, out you come, we've got you.' Then, not bothering to use his whistle, he gave a roar: 'Right Bob, here they are.' Seconds later

Hedges joined us, as the two men were getting to their feet. Both were new faces to us.

Bob turned to George. 'Have a look and see where they've dumped their gear.' It did not take much finding, for it was pushed in under the gorse next to where the men had been lying – two sacks. George picked up one and gave it a shake. Out fell a net. He did the same to the other and out tumbled twenty-three partridges, all with their necks broken – at least two coveys. I thought for a moment that Bob was going to wade into the two men, but he managed to restrain himself. We marched them off to Tom Rose's house and phoned the police. They were most interested in our capture and recognised them at once. Apparently the police had been looking for them. They were wanted for questioning, in connection with breaking and entering two nights previously. It came out in court that both men had quite a record of night poaching and were sentenced to six months' hard labour. What led to their discovery was not given in evidence!

I had been down to Hay Mill to check that a new gate and fence that had recently been erected, where the meadow ran out to a point and joined the pond just short of the mill, were still intact. It was truly a barbed wire entanglement – not long, only about twenty yards from the virtually impregnable boundary hedge to the gate, the other side of which were the rushes and bottomless mud that flanked the pond – but an excellent deterrent to any law breakers trying to gain entry. The wire had twice been cut by vandals and had only been replaced that morning. The locked gate was framed with tight loops of barbed wire. The only way through was either with wire cutters or a key.

All had been well. It was a lovely warm evening and I cycled quietly along the road beside Nobb's Crook, a large arable field to the west of Lamas Wood. The light was starting to go, but I just caught a glimpse of someone, about 150 yards ahead, turning into the gateway. I speeded up but when I reached the gate there was no sign of anyone. Knowing that there were two partridge nests not far along the inside of the hedge to the right, I quickly entered the field. There, sauntering along, some twenty yards away, was a youngster whom I adjudged to be about thirteen or fourteen years old.

'Hey, where do you think you're going?'

'Only to 'ave a pee, mister.'

'That's a likely story. You wouldn't have bothered to step off the road.' He grinned and walked back towards me, oversized shorts flapping below his knees, the braces holding them up kept together with bits of string. A dirty collarless shirt, also several sizes too big, and an old cloth cap, set at a jaunty angle and covering a mop of red hair, completed his regalia. His face was a mass of freckles and his green eyes had a mischievous sparkle.

'You goin' ter clobber me, mister?' Then, with a grin from ear to ear, 'That's if you can bleedin' catch me.'

'No, not unless you've done something to deserve it. Where do you come from?'

He made a gesture with his thumb over his shoulder. 'From th' camp.' I knew there was a party of didicoys who had set up camp, just beyond the estate boundary, on some rough ground.

'What's your name?'

'Aint got a proper 'un. They calls me Slither, 'cos I can get fru anytin.'

'Well, master Slither, what were you doing creeping along inside the hedge?'

He looked down at his bare feet for a moment and then, still grinning: 'I were goin' ter watch th' sport.'

'Sport?'

'Yeah, two blokes 'ave just gorn across th' field wid a

long net an' a dog. I asks un what they were doin' an' them said they were goin' fishin'. Must think I'm bleedin' daft.'

I questioned him further and it sounded from his description, particularly that of the dog, as though one of the men could be Jim Sloane – although it was some distance from where he normally operated. I felt in my pocket and found a shilling. 'Here you are Slither, thanks and if I catch you up to no good I'll . . .'

He interrupted. 'Cor, thanks mister, yer a toff. Don't you worry, I don't do th' dirt on me friends.' He spat on the coin and put it in his pocket, as he headed off jauntily down the road.

Home was the nearest place to raise the alarm and in a little over five minutes I was in the car with father, heading for Lynch Hill. We were lucky. We met George at the top of Cocksherd Meadow and picked him up. Luck was again with us, for Bob had just returned to his cottage. The four of us made haste to Lamas Wood. George and I were to go to the far end and double back towards the middle, keeping just inside the hedge. Bob and father would work their way up from the lane. We felt fairly confident that Jim Sloane, if indeed it was he and his crony, would have staked the net out at the top end of the wood, adjacent to a piece of late sown spring barley. There was a tremendous number of rabbits doing untold damage here – so much so that Jummy Young and his helpers were to start putting up a new wire netting fence the following day.

George and I reached the far end and quietly started to make our way back, just inside the wood. I was in front when suddenly, only yards ahead, I heard someone clear their throat. Just at this point there was a patch of snowberry, so enjoyed by pheasants, running out to the edge. I signalled to George to stay where he was and inched forward. At last I was able to peer through and could just discern the shape of a man, his arm out from his side, obviously resting on the net. As I watched he ran forward, along where I assumed the net was set, with a stick in his hand. He reappeared within seconds carrying a rabbit, which he threw on the ground. It was only then that I noticed a heap which I felt must be rabbits. A dog loomed up through the dimse. I felt certain that it was a lurcher, from the silhouette. I distinctly heard, 'Good girl, Rose. Come here.' The dog immediately disappeared into the gloom. There was no doubt that the figure I could see was Jim Sloane!

I wriggled back to George and, whispering quickly, put him in the picture. We crawled forward oh so quietly, literally inch by inch, until at last we were out on the edge of the wood. Slowly we stood up, hardly daring to breathe. The light was fading rapidly, but we could make out a figure some 6–8 yards ahead. Then he disappeared, and we could hear him running off down the net. This was our chance. We moved silently forward and pushed back into the shadow of the hedge by the start of the net. About a minute later a man appeared and we could just see that he was carrying two rabbits, which he added to the pile on the ground. George took one step forward and grabbed him by the collar.

'Got you!' The man gasped, then yelled 'run for it.' We heard pandemonium break out some 80–100 yards further down the edge of the wood. George shone his torch on his captive – it was Sloane. At that moment a man came tearing up the field. George made a dive for him. An international rugby footballer could not have handed anyone off better. George crashed to the ground, having received a hand full in the face. Cursing, he staggered to his feet as the fleeing man disappeared into the darkness. Just then father and Bob joined us; I turned my torch onto

George, his eyes were running and blood was pouring from his nose. Without a word he set off in the direction the man had taken. I joined him. We were running across Hay Mill Meadow when I stopped.

'George, this is daft, running about in the dark. We haven't a hope.'

'I know, but there's a slight chance. He may have headed towards the mill, we just might corner him by the new fence.'

We jogged on until we were about 200 yards from our objective, then stopped. I could have sworn I heard something. 'Listen!' We stood silently for a few moments and then distinctly heard cursing, followed by a cry for help.

George rubbed his hands together. 'I think chummy's tried to get through the fence and has got hung up. Come on, let's go.' We didn't run; the flood of obscenities told us that there was no need to hurry. When we reached the fence we found our man well and truly caught up in the barbed wire. He'd managed to get one leg through and had then become completely ensnared along his back, front and down both legs. George kept the beam from his torch on the man's face. 'That'll teach you to go poaching and hitting people in the face.' He drew back his foot to deliver what would have been a vicious kick, but I stopped him. The man cursed and then said, 'I haven't been poaching.'

'Oh, well then, you're not the man we're looking for.' George turned away. 'Come on, let's look back up the hedge. I tell you what, the old bull won't half sharpen his horns on matey when he finds him.'

'Hey, wait a minute and what's this about a bull?'

'You got yourself hung up like that and, as you aren't the poacher I'm looking for, you're no interest to me. Don't worry about the bull, he's never killed anyone yet, only put two or three people in hospital with broken bones.'

116

We continued to walk away and were followed by a flood of profanities. George stopped and then, to my surprise, gave a very fair imitation of a bull, which was followed immediately by yells for help from the entrapped man. We returned slowly, George stopping to render a couple more bull impersonations, before putting on an excellent act pretending that he was driving off the bull! When we got back to the fence it was a very shattered and subdued man that we found. Without being asked he admitted that he had been poaching with Jim Sloane and would we please release him. This we were unable to do until I had been to the mill and borrowed a pair of wire cutters. Having freed him, George searched him and was fortunate to find that he had a driving licence, which I pocketed. His name was Harry Jennings and he swore he'd never been poaching before.

We set off across the field, heading for Lamas Wood. About halfway across George stopped and said he'd catch us up in a few minutes. We hadn't gone very far when suddenly there was a bellow of a very angry bull not far behind us. Not expecting it, I jumped nearly as much as the terrified Jennings. George was enjoying himself! He came running up. 'Quick, let's get out of here before that old devil finds us.' He grabbed our prisoner's arm and started to run to the wood. When we were through the fence the latter collapsed onto the ground, nearly in tears. I shone my torch on him.

'I've never done anything outside the law before and if I get through this night alive I never will again. That flaming bull, they're the one thing that terrifies the life out of me!'

I thought for a moment George was going to give the game away, but he held his peace. We made our way through to Lynch Hill Lane. The police had arrived and, as we drew near, I could see Sergeant Garrett and a constable. I gave the driving licence to the former. The two miscreants had netted nearly forty rabbits.

It proved to be Jim Sloane's swan song. He already had a number of outstanding fines to pay, and the very heavy one levied when he appeared in court for his most recent offence was the final nail. He was given two weeks to pay; if not, the chairman told him, he would go to jail. Two days before his time was up Jim appeared at the estate office, together with Rose and his ferrets. It was the only time I ever felt sorry for him. He begged father to take Rose; he said that she wouldn't be looked after while he was away. Father agreed to find her a home; the ferrets were no problem. Next day father took a fretful and obviously pining Rose up to Oxfordshire, to Tommy Letherbridge, a warrener he knew well. She quickly became one of the family and Tommy said she was the most intelligent lurcher he'd ever come across. She lived to a ripe old age and spent many happy years rabbiting, in some form or other, most days of the week.

The last I saw of Jim Sloane was walking out of the estate office, tears silently running down his cheeks. I think at that moment, had I had the money, I would have paid his fine and reunited him with his beloved Rose. If only I hadn't seen Slither . . . but then life is full of IFs!

It must have been an evening in early November 1937, because I was driving father's car and had only had a licence a few months, when I was taking a message up to Cyril Barker at the Leas Farm. I came down the hill to Cocksherd Wood and saw two bikes leaning up against the gate leading into the grass field that adjoined the wood. This immediately made me suspicious and I decided to investigate further on my return journey. Ten minutes later I coasted down the hill, engine and lights off. There was a full moon and it was nearly as light as day. I stopped by the bikes and got out of the car. I could hear the cock pheasants in the covert kicking up a racket – something was very definitely wrong. I raced home and told father. Within minutes we were on our way to Lynch Hill. It was 6.00 p.m. and we caught Bob and George just as they were off out on their rounds. Saturday was the first major shoot of the season and they would be patrolling their beat until 2.00 or 3.00 a.m. After the minimum of explanation from father, they climbed into the car and we were off. As we turned into Crow Piece Lane father switched off the lights. He stopped at the end of Cocksherd Wood and Bob jumped out and headed off along the ride, while we continued on down the road to where I'd seen the bikes. George was to wait at the main entrance from the meadow into the wood. I was to remain by the bikes and father would return to the top of the wood where we had dropped Bob.

I had only taken up my position for a few minutes when it occurred to me that it might be wise to move the bikes. Across the road was a gate into a field of swedes, so I pushed them some thirty yards out into this and laid them down – and, as an afterthought and for good measure, I let the air out of the tyres! I returned to my post, concealing myself in the shadows. It was only a short while before the comparative quiet of the evening was broken by a positive uproar coming from where George was stationed. Apparently he had gone to stop two men, fleeing from Bob, and both had turned and attacked him. The speed with which they had done this had caught him by surprise and they had knocked him over. However, he was able to grab one by the ankle and pull him down. The other came racing along the field to where I was waiting and jumped over the gate. As he landed he gave a curse on seeing that the bikes had gone and turned to run up the road. I only had to stick a foot out to bring him crashing down. Fortunately he was well and truly winded, for he was a big man, so I leapt on him, twisting his arm behind his back and, at the same time, blowing my whistle to let father know he was wanted.

It was all over in minutes. The two poachers were bundled into the back of the car, the two gamekeepers literally sitting on top of them until we reached the police station a mile and a half away. The poachers had a powerful airgun and a bag full of pheasants, the gun fitted with a torch, in the same way as my ratting gun. All the pheasants had been shot through the head, mostly through the eye, with little darts made to fire from air-guns. While both men were unknown to us, they were no strangers to the police and had numerous convictions for petty theft and assault, as well as poaching. They were in trouble for, apart from night poaching, they had threatened Bob with the gun and attacked George. When they were eventually released, they asked where their bikes were. I told them and added that had they not left them on the road it was unlikely they would have been discovered. I often wondered if it crossed their minds when, some weeks later and they'd both received stiffish jail sentences, that a few moments spent hiding their bikes would have been time well spent!

Pound Cottage, 1936. Long since flattened and replaced by numerous houses.

Red Poll bull, Bredfield Nathan, 1926.

Burnham Grove Estate staff in 1924.
Back row, left to right: — Johnson; Fred Goodchild; Harry Jaycock; Jack Keen; ? ; ? ; believed to be 'Gunner' Stingmore; ? ; ? ; Arthur Goodchild; — Goodchild (snr); believed to be 'Froggy' Hawkins; ? .
Middle row: Harry Stanley; ? ; Jack Herbert; Jim Brookling; Charlie Coxhead; Fred Simpson; Tom Brookling; Bert Stanley; Alf Small; Sam House; Billy Taplin; ? ; — Knight.

Front row: — Jaycock (Old Jay); Bill Herbert; Charlie Davis; Bert Ridley; Ernest Taplin; Tom Rose (farm foreman); Cecil Twist (land agent and farm manager); — Bangell; — Pudifoot; — Miles; — Whitworth; — Murkett.

The author's parents, Cecil and Mabel Twist, 1939.

The author's father, sitting on the former's 'roadster', nursing his pet badger, Joanie, 1923.

The first pigeon. Left to right: *The author, Bridget, Cecil Twist, the author's brother, Ralph, holding Wendy. Springer spaniel, Zena, in front.*

The fire at Britwell, 1928. Smouldering hay being thrown off the rick after the fire was under control.

The author and his brother beside the bull-nosed Morris Cowley, that regularly went up and down Peppercombe in North Devon, on a road which for decades was marked 'Impassable for motors'.

Burnham Grove, the home of Mr and Mrs Edward Clifton-Brown.

Group at Rhona Clifton-Brown's wedding on 23 June, 1927. The bride and the groom, Major Weston Cracroft-Amcotts, are in the centre. Edward Clifton-Brown is to the left of the bride. To the right of the groom are 'Dolly' Clifton-Brown, Mrs E. Cracroft-Amcotts and Anthony Clifton-Brown, the bride's younger brother.

*Fred Goodchild with a
Berkshire sow and her
litter.*

*Whitewyck Garnet, the
author's pet Dexter, aged
about two and a half years.*

H.R.H. the Prince of Wales (left) having just presented to C. J. Twist (3rd from left) the Prince of Wales Gold Cup for the Supreme Champion pen of sheep, Smithfield Show, held in the Agricultural Hall, Islington, 1930.

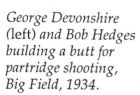

George Devonshire (left) and Bob Hedges building a butt for partridge shooting, Big Field, 1934.

Part of the Burnham flock being folded over red clover. Shepherds can be seen in the background about to set hurdles for another section.

The sheep sheds. Harry Wadman with shearling rams in the foreground.

Some of the show team, 1938, well controlled by two working Old English Sheepdogs – now better known for advertising paint!

Two-year-old ewes in Gray's Meadow. A four-wheeled wagon and shepherd's hut can be seen in the valley.

Fred Simpson with a Devon bullock. Reserve Supreme Champion at Slough Fatstock Show, 1938. A thatched corn rick and carts can be seen in the background.

Jerry, the author's pet grey squirrel, enjoying his bottle.

Berkshire sow, Burnham Griqua Lass. Winner of numerous first prizes and Supreme and Breed Championships in 1933.

The fox cubs at about ten months. Brandy 'combing' C.T.'s hair for Brylcreem, while Dom looks on.

The Champion Pen of eight Hampshire Down ram lambs, from the Burnham flock, Salisbury Sheep Fair, 1935. It often took a day – or longer – to persuade all eight to stand in line!

Harry Wadman, trimming one of the flock in preparation for a show.

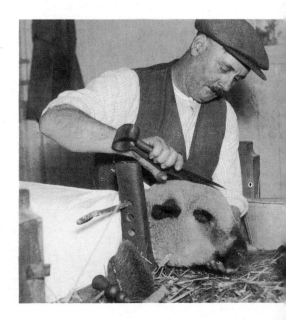

The fruits of their labours: trophies won on the Burnham Estate in 1937. Back row, left to right: Harry Wadman (head shepherd), Fred Goodchild (head pigman), Mr Brooks (head gardener), C. J. Twist (agent), Tom Rose (foreman), Fred Simpson (head herdsman). Front row: Les Holloway (herdsman), Cyril Barker (head poultryman), Albert Goodchild (pigman), Philip Wadman (shepherd).

Changing times, 1939. Wattle hurdles have replaced the 'open' split willow hurdles. The lambs in the foreground have the pick of the grazing; the ewes clean up in the background.

The Trinity House boat leaving the Wolfe Lighthouse on 24 August 1939. This was just before coming alongside the boat from which the author was fishing and advising an immediate return to port as war was imminent.

The author by a mega crop of narrow stem kale in 1941, a year after taking over the management of the Round Hill Estate near Aylesbury, Bucks.

10

Grass Snakes and Squirrels

'WOULD you like a hard-boiled egg with your tea? They're freshly laid today.' Mother looked enquiringly at our guest.

'Yes please.'

'How many?'

'Would six be all right?' Mother nearly managed to hide her surprise, but not quite, as she turned to leave the room. Our guest was tall, fair haired and had arrived in a somewhat unusual way via a blazing aircraft!

I had accompanied father to Big Field, to look at a rick of clover hay which had been opened that morning and a wagon load trussed, that is, cut out with a hay knife in blocks some 4 feet by 2 feet 9 inches and, when the strings were pulled tight, making a truss about 21 inches deep. When doing this it was customary to cut the rick out in 'benches'. This particular rick, after the load had been tied, had a bench the width of the rick, about 10 feet deep and 6 or 7 feet from the ground. Father had just completed his inspection when we heard an aeroplane approaching, not a common occurrence in those days. It was heading straight towards us.

As it came into view it was obvious that the pilot had problems, for the engine was spluttering and smoke was starting to pour from it. Finally it cut out as flames began to appear and in seconds the fuselage became alight. I could see it had two cockpits. As we watched, the pilot climbed out and sat sideways on the plane. It crossed The Belt, glided straight over the rick, clearing it by some 50 feet and, at exactly that moment, the pilot jumped, landing on the 'bench' and rolling over before sliding off. The plane carried on for a further 50–60 yards before hitting the ground; immediately it was engulfed in a sheet of flame. Being largely constructed of wood and fabric it did not burn for many minutes. The pilot gave us a cheery wave and sauntered over. He introduced himself, remarked that that had been a near one and that he was b. . . .y ravenous. Had we a telephone? Father told him that we had and that he had better come back to the house. We climbed into the bull-nosed Morris and were away – and our passenger never stopped talking! Apparently he had taken off from Hendon, one of the two small airfields close to London (the other was at Croydon) for 'a bit of a jolly' and to see if he had done a good job tuning the engine. When father remarked that it appeared he hadn't been too successful, our guest roared with laughter and wholeheartedly agreed. He put a call through to Hendon and told us that a chum was motoring over to pick him up. That was when mother asked him if he'd like to join us for tea.

I sat positively goggle-eyed as the intrepid pilot made short work of six eggs. These he followed with a plate of sandwiches – it was not a small one – numerous slices of bread and butter piled high with comb honey and finally several large slices of cake. Bob, as he asked us to call him (I don't think I ever did hear his surname), was charming, totally gregarious and, apparently, unaffected by the afternoon's events. Having at last finished his tea, which he assured mother was the best he'd ever had, absolutely scrumptious, he asked if he could see some of my pets, about which I had been telling him. Father made his adieus and said he had work to do. Bob thanked mother profusely and, to my amusement and her amazement, gave her a smacking kiss, a slap on the bottom and assured her that she was a doll! We did the grand tour of the menagerie, Bob showing a great interest in everything, particularly my snakes. He also displayed a keen interest in the stubble field adjoining the garden – in fact, he walked out and studied it quite carefully. Yes, he told me, one could land a small plane there, but it would be tight.

A week later we were just sitting down to have after-noon tea when I heard an aeroplane. The dining room looked out across the field. Seconds later a plane skimmed over the far hedge, touched down almost immediately and came to a halt close to the gate into the garden. The pilot jumped out, took some things out of the cockpit and headed for the house. It was Bob. He did not stand on ceremony, just walked straight in and announced that he had come to tea. He had also come to say thank you again and to bring mother a huge sheaf of flowers, a vast box of chocolates and two bottles of malt whisky for father. He ate an enormous tea, but no eggs this time, chatted away non-stop and then announced that he must be off. He cornered mother, in spite of her trying to dodge him around the dining room table, gave her another kiss and assured her she was a gorgeous creature. Mother tried to appear affronted, but it was the only time I can remember her positively simpering! He asked me if I'd like 'to go up for a spin' and seemed mildly hurt when I declined, then got in his plane, revved up and was gone. I watched him clear the hedge by about two feet and was very glad I had not accepted his offer.

Life was never dull. Quite apart from seasonal happen-ings, there always seemed to be something more to add to a full and interesting life. Snakes were not a rarity on the estate and one place in particular, in the main woods, seemed to be the home of grass snakes. Ralph and I not infrequently, on very hot days, would go in search of these, but we were seldom successful. Snakes on the whole are timid creatures. However quiet one is they seem to sense one's presence and quickly slide away. However, on a few occasions we caught one napping and were able to capture it. We did this by pinning it gently to the ground with a forked stick, just behind the head, so that we could pick it up without any fear of being bitten – not that a grass snake is poisonous, but it can give one quite a tweak. We always released any we caught. Sometimes, if they seemed big, we would measure them. I caught one once on my own, it was a 'biggie' and with great difficulty I measured it. If you have never tried holding a scared and obdurate snake, while at the same time trying to ascertain its length with the aid of a tape measure, I can assure you it is no easy task. However, I eventually succeeded; it was 3 feet 4 inches, give or take a wriggle or two! I was tempted to keep it, but already having a number at home, I turned it loose.

Our pet snakes came from the London Zoo. We were very friendly with Reg Lanworn, the head keeper at the reptile house. Grass snakes and Italian tessilated snakes were used for feeding some of the larger varieties. If there was a particularly nice specimen in a food consignment, Reg Lanworn would put it in a tin, half-filled with damp moss, make some air holes and post it to us. Alternatively he would keep it for us until next we called to see him. We were very regular visitors and, in winter time, often spent two Sundays a month at the zoo. In those days one could only get in on a Sunday with Fellows' tickets. Edward Clifton-Brown was a Fellow, but had no real interest in the zoo and used to give all fifty of his tickets to father.

One morning, just before we left for school (it was after the debacle of sending us to boarding school), a parcel arrived addressed to Ralph and me. We knew the hand writing, so immediately took it to what had been our nursery-cum-school room and had subsequently become a section of our 'zoo'. It was filled with glass cases for reptiles, fish tanks and cages for small rodents. Time was short. We hastily prepared an old accumulator tank and, with difficulty, deposited our new acquisition in it – a big grass snake. It had travelled well and was very active! A piece of cardboard with air holes in it was placed on the top, with a large book on each corner. When we returned that evening from school, our new pet was foremost in our minds. To our horror and amazement it was missing. So was one of the books! Subsequently, we discovered that the maid we had at the time wanted something to read and, with all the hundreds of books there were in the house to choose from, she had to go and take one off the snake's tank. We searched frantically, but to no avail.

To get into the room one had to come through the kitchen. There was another door leading to the main part of the house which, for various reasons, but chiefly because of mother's phobia about snakes, was kept closed. A long passage led through the house from the kitchen to the front hall, from which one entered the drawing room. Ralph and I were at a loss to know where our new pet could have gone and where to look next. Suddenly the most terrible and continuous screams filled the house: they were coming from the direction of the drawing room. We raced through to find mother in hysterics. It was some minutes before we could find out what the trouble was and why she was so upset. Apparently, she had gone to her bureau, pulled open a drawer and reached in, without looking, for whatever it was she wanted. Our missing snake had slithered out over her hand, having made its way through the house and somehow climbed up the back of the bureau and into the drawer. It took us a good ten minutes to find our escapee, but it took an awful lot longer to live down the episode! With hindsight, I think it took great magnanimity on mother's part that all our snakes and other reptiles were not banished from the house. There was one sad repercussion. Reg Lanworn was asked by father not to send us any more snakes through the post.

Around this time several sightings of a monster grass snake were reported both by two of the woodmen and an under-gamekeeper, who assisted Bill Yeoman on his beat. On each occasion it had been seen in close proximity to the favoured haunt of grass snakes, in the main woods. Its estimated length varied from 6–8 feet. Father was very sceptical about this and said that if there was indeed a snake of this size in the woods, then it must be someone's pet python, or some such similar species, that had escaped. I spent many hours looking for this phenomenon, but although it was seen twice in one week I had no luck. Father gave strict instructions that it was not to be killed, but alas to no avail. One afternoon Bill Yeoman arrived at the estate office, to say that he had been walking along the grass ride near 'snake corner' when a monster snake had reared up out of the undergrowth, hissing. He thought it was going to attack him, so he had shot its head off. Father was furious, for Yeoman's story was obviously a fabrication: grass snakes simply don't rear up when disturbed, they scarper as quickly as possible! Father went up to the woods to have a look at the remains. As he had anticipated, it wasn't quite the monster that had been reported; nevertheless, the headless body measured 4 feet 7 inches, so it was probably about 5 feet. Undoubtedly, according to John Norman of the Natural History Museum, it could have been a record British grass snake.

One reptile that mother was not averse to was Sally, a yellow and black salamander, a type of American lizard, that lived in her glass case at the end of the central table in the estate office. When we were at London Zoo we spent much time with Reg Lanworn, in the area behind the display cases, 'playing' with various snakes, lizards and baby crocodiles. One day Ralph found Sally wandering around. She had inadvertently been removed from her enclosure, which she shared with numerous others of her kind, with some potted palms and other greenery. Ralph asked if he could keep her and was told yes; apparently the zoo had an abundance of salamanders.

Sally became internationally known. People came from all over the world to buy pedigree stock from the herds and flocks on the estate. Invariably they ended up in the estate office and would take an interest in Sally. A salamander's tongue is reasonably long. Father would take a mealworm or two, her staple diet, from a tin, put them on the table and lift Sally out, putting her down a few inches away from them. She would remain motionless, except for her blinking eyes, then suddenly, almost faster than the eye could follow, her tongue would shoot out and she would be happily munching a worm. A party of Japanese, who came annually to buy Berkshire pigs, were so taken with her that one year they brought her a present of some mealworms!

Occasionally our visits to the zoo would be highlighted by some incident other than 'playing' with snakes, baby crocodiles and alligators. On one occasion Reg Lanworn said he'd get George to show us his teeth. George was a big alligator, some 9–10 feet long, who was said to be about a hundred years old. We had seen George's teeth many times before, but it had become almost a ritual every time we visited the reptile house. Lanworn would take one of the long thick bamboos, with a piece of L-shaped metal on the end, used for lifting poisonous snakes, lean out over the pool and gently tap George on the nose. Sometimes he would pay no attention, but normally he would respond magnificently, much to my delight. He would open his cavern of a mouth and give a positively hair-raising, blood-curdling roar, before subsiding to the bottom of his pool.

Out of sight to the normal visitor, a catwalk extended out over the pool; its purpose was to enable the keepers to pull out the centre plug and change the water. The walk had iron railings either side and a gate at the end, and was an area that was definitely out of bounds. However, one day when I was in a particularly boisterous frame of mind – I was about eleven or twelve at the time – I decided it would be an excellent vantage point from which to view George. I deliberately lagged behind as we approached and, as I saw the bamboo being pushed out over the pool, I ran up the ladder, rushed along the catwalk and hit the gate with a bang. It wasn't fastened! Somehow I grabbed the rails either side of the gate as I shot forward; fortunately my feet remained on the walk, as I hung only a few feet above George's open mouth. I was able to confirm that his teeth were even more massive than they appeared from the side line and his breath was foul! Father, at a speed that appeared to be quicker than light, was there to pull me back, but he needn't have hurried. I had absolutely no intention of going anywhere, least of all forwards! That little adventure caused me to be debarred from the next three visits to the zoo, in spite of my assurance that there would be no repetition. I had seen all I wanted to of George's teeth.

Another house pet, rather more popular on the whole than the snakes, and one that always caused great interest and much amusement was my grey squirrel – Jerry. When very young, quite unintentionally, he might have cost me my life. One evening I was in Bangle's Spinney, the next covert to Lamas Wood, checking on a number of birds' nests that I was recording, when suddenly there was a shot the other side of a large clump of laurel not twenty yards away. It was George Devonshire, and he'd shot a squirrel. They were, and indeed still are in spite of their endearing appearance, vermin when in their natural surroundings. I chatted to George for a few minutes before he said, 'Well I must be off, but I better put a shot through the nest before I go.' With that he fired at the squirrel's dray, in the top of the tall larch tree under which we stood. I went back to checking the nest I had been about to look at when George had fired.

Ten minutes later I was walking back under the same larch and happened to look up. There, very close to the top, was a baby squirrel clinging precariously to a branch. As I watched it swung round underneath and, with much difficulty, scrambled back to safety. Obviously it had to be rescued. I looked at the tree. The first 15–20 feet were completely devoid of branches. Fortunately I had my climbing irons with me, so I strapped them on. As I neared the top, the tree started to sway in the most alarming fashion. I stopped some four feet from my goal. Was it worth going any further? I looked at the pathetic little creature clinging to the branch, terrified and eyeing me with uncertainty. The answer had to be 'yes'. Slowly I eased myself upwards, the trunk becoming almost non-existent, but at last I was there. I reached out – I dared not look down – but the baby squirrel never moved; he seemed paralysed with fear. At last my fingers touched him and, as my hand closed around him, he buried his teeth in my thumb! My sudden and surprised reaction to this unfriendly behaviour caused me to lose my hold, which at the best could only have been said to be precarious, and with a yell I started to fall. I retained my hold on Jerry, but then he wasn't letting go anyway!

I seemed to be falling for ever, although I only dropped some 15–20 feet before I was brought up short astride a branch – not a pleasant experience! With my free arm I grasped the trunk as I got my breath back and tried to steady my nerves. One good thing was that I had neither squeezed Jerry nor let him go, although I doubt at the time the latter was possible, for he still clung firmly to my thumb! I don't know how long it was before I realised I couldn't remain where I was indefinitely and somehow I had to get down. Not an easy task wearing climbing irons and with only one available hand. It took me several minutes to clear the branch I was astride; from then on it was a slow, difficult and hazardous descent, which wasn't helped by Jerry suddenly changing his hold to a finger.

At last I felt the ground under my feet and sat down, leaning against the tree, exhausted and not a little shaken. I inspected the orphan I had rescued. Yes, he was definitely worth it. Gently I stroked the top of his head; to my relief he withdrew his teeth from my finger. He never made any attempt to bite me, or anyone else, again. It was only at this juncture that I took stock of myself. My shirt was torn, the inside of my trouser legs could only be described as shredded, my arms and legs were grazed in a dozen different places. I put my free hand up to my face – it came away sticky with blood. I wiped my fingers on my trousers (they were a write-off anyway) and started to stroke the top of Jerry's head again. He blinked and eyed me, but showed no trace of fear or animosity. With some difficulty I unstrapped my climbing irons and got to my feet. It was only then I realised how scratched and bruised I was. It began to dawn on me that perhaps I had been very lucky to get away with such superficial damage.

I arrived home some fifteen minutes later. Luck was no longer with me. I opened the six foot high boarded back gate to come face to face with mother. It was immediately apparent that my appearance did not please her! Briefly, I told her the cause, which led to my having to suffer a terse harangue on my stupidity and did I not realise I could have broken my neck? When it was over I introduced her to Jerry. Her attitude softened immediately and she had to admit that he was rather sweet. I took him into the house, put him in a cage in our zoo room and went to clean up. Half an hour later Jerry was sitting on the table happily taking milk from a fountain pen filler. He quickly graduated to a baby's bottle. He truly was a big baby, because he loved this and would sit clasping it with his front feet as he contentedly sucked away. It was months before I could separate him from his beloved bottle.

Jerry's early efforts at jumping were as hair-raising as they were amusing. The first time he launched himself was one day after I had given him his evening feed, on the table in the middle of the room. He was terribly inquisitive. Having run round carefully examining everything, he came and sat on the edge of the table facing me: I stood several feet away. Suddenly he started to jump up and down on the same spot, chattering. Without further warning he launched himself in my direction. Fortunately my reflexes were quick as he, virtually, came flying over my head. I just managed to catch him – had I not done so he would have crash-landed in an aquarium! He had an alarming tendency to overpower his jumps for several weeks, before finally sorting out his act. Once he had mastered this he would, particularly in the evening, set off around the room in a positive 'wall of death'. He never fell and never broke anything. Jerry would sit happily on my shoulder for hours if allowed, either in or outside the house, and never attempt to take off.

I have no idea what the normal life span of a grey squirrel is but, alas, poor Jerry never made it to his second birthday. He died really from a surfeit of love and good intentions, for his demise was due to a fatty heart. Everyone loved him and, in spite of my protests, fed him 'a little something' every time they visited him. He was a 'guts' and would never say no, but he was, without doubt, the most endearing pet I ever had. And for many years I carried a scar on my thumb as a memento of our first meeting!

While Jerry was indeed a most beloved pet, in the wild grey squirrels can be a terrible pest, under certain circumstances. They will do untold damage to birds' nests: they love eggs and fruit is also a great favourite, particularly cherries. Squirrels became such a pest in the orchards at Burnham that, eventually, a bounty of 6d each was placed on their tails. In 1939 the bird scarers – these were men walking round and round the orchards, with old muzzleloaders, black powder and shot, frightening off and indeed shooting the birds – accounted for 681 grey squirrels. It was ironical that Edward Clifton-Brown should be paying a bounty on them in 1939, when he was the person who in 1922/3 had introduced them to the area and released a large number into his woods. Some years previously he had obtained several pairs which had increased and multiplied rapidly in captivity and, in spite of a number of people advising him to the contrary, he had turned them loose. The indigenous red squirrels, so attractive and present in goodly numbers in the twenties throughout the district, became conspicuous by their absence by the mid to late thirties. There are those who, today, try to claim that the grey squirrels were not responsible for the demise of the reds, but their disappearance, shortly after the greys infiltrated an area, lends little credence to this theory. Be that as it may, vermin I know they are and a terrible pest on shoots, or in orchards, but since my association with Jerry I have never shot one. They are welcome visitors to my Suffolk garden, adjacent to a large wood where annually they hold a squirrel and fox shoot – I think with little success. One regular visitor to my garden, from these woods, is minus his tail. He is far shyer than his companions, but he shouldn't be, for he carries no bounty!

11

Threshing

T HE majority of farmers relied on contractors to carry out their threshing. In the twenties and early thirties the threshing machines would nearly always be driven and moved around by a steam engine. Magnificent sources of power, they were normally polished and cared for by the driver with a pride, and indeed love, which would appear totally foreign these days. However, the estate had its own 'tackle' – a Marshall thresher and, eventually, a straw tier, moved and driven by a wonderful old tractor, the traction of the two rear wheels being independently controlled by long hand levers reaching back either side of the steering wheel. It had tremendous power for those days and was fitted with a long steel hawser and drum which enabled it to winch the thresher through the heaviest of mud should it become stuck.

Jack Keen was always in charge of the threshing gang. It was his task, aided by Tom Brookling, to set up the 'drum' beside the rick that was to be threshed. It had to be just the right distance away so that the side flaps, which would be up when travelling, let down to cover the space between the machine and the rick. It was most important that the former was level and the tractor had to be aligned exactly so that there was no fear that the big drive belt would fly off at speed. There was no question of any belt guards: both the main drive and the numerous belts on the thresher were devoid of any covering. Nor was there any protection to stop the 'feeder', or indeed anyone else in the gang, from falling into the drum. The 'feeder' stood in a small well in the deck and sheaves of corn were passed to him, the strings already cut. These he would 'dribble' into the revolving drum in front of where he stood. Feeding was a job that required constant attention to obtain the maximum throughput. If the flow was insufficient then the machine would not be threshing to its full capacity, too much and one could clog the drum. This was not difficult to do if one dropped in several sheaves quickly without shaking them out; belts would start to slip and everything would come to a halt.

More than one farm worker lost a hand, or indeed an arm, trying to clear a blockage before the ganger had cut the power. This, ironically, would only happen if the 'feeder' was successful, for the drum would suddenly roar into action and start to rotate, drawing anything within its orbit down inside. Before the first time I had a go as 'feeder' I was given a stern warning by Jack. To emphasise the danger, he told me, in the most graphic terms, of a man he had once worked with who lost both arms up to the elbows whilst trying to clear a clogged drum before the power had been stopped. I took good heed of his warning.

Normally there would be three men pitching the sheaves off the rick. One stood on the edge forking them on to the 'cutter', dropping them in front of the latter who would lean forward, pick up the sheaf, turning it as he did so with the knot between his forefinger and thumb so that, all in one movement, he passed the sheaf to the 'feeder', cutting the string right by the knot. Usually a razor sharp curved knife, like a pruning tool, was used for this purpose and it was not too difficult to take a lump out of oneself, particularly at the end of a long tiring day. Cutting the bonds was a back-breaking job; in fact there was only one worse – being on the chaff and cavings, the place were every boy started. I was no exception. The chaff was blown out at the side of the thresher and, if the corn had been carried without being exposed to a lot of wet weather, this was bagged off in the case of oats and wheat. Subsequently it would be used for feeding to the work horses with their bran and oats, or in the case of oat chaff it would be mixed with pulped or, more accurately, sliced mangolds for dry stock (cattle), together with a little barley meal. The cavings, which came out under the straw shakers, were the rubbish. Clearing this and looking after the chaff entailed working all day in choking dust. One had literally to go in under the end of the thresher to move the cavings – a filthy job, even when the crop had been harvested in perfect conditions and the field was comparatively free from weeds. By the end of the day one looked like a fugitive from the Black and White Minstrel Show!

I had to take my turn, for father was adamant that if, in the future, I intended managing an agricultural estate then I should have extensive practical experience, as well as theoretical knowledge. He was right, of course. But after my first day clearing the cavings and chaff, I would have taken a lot of convincing as to his wisdom! Worst of all was barley: if ill prepared the awns got down inside one's clothes, and it seemed as though they were alive, there was just no stopping them. By the end of the day it was rather like wearing a hair shirt!

The first time I was on the chaff and cavings we were to thresh a small rick of barley. It was in the autumn, and it was a very warm day, so I had on an open-necked shirt, sleeves rolled up, and nothing on my head. As we waited for Jack to start the drum going, old Gunner eyed me thoughtfully, spat with deliberation and remarked, 'You'm larn.' I did. The next time I had a cap on, my shirt buttoned up and a 'sweat rag' around my neck. Further, much to the gang's amusement, I had a large handkerchief tied around my face, covering my mouth and nose, and tucked down inside the collar of my shirt. By so doing I felt sure I would avoid inhaling several pounds of black dust. Sometimes, if things were going very well, extra help would be needed on the cavings and chaff, but never was a second boy sent to assist. There was a general belief, in those days, that when you set a boy at a job you had a boy. If you put two to work together you had half a boy and three would mean you had no boy at all! It was on the whole a good maxim and one that I proved to be true in later years.

Really the threshing machine was far more sophisticated than the modern combine. Certainly the latter devours vast quantities of standing corn in a day, retains the grain in a hopper until offloaded into a trailer and spews out straw, cavings and chaff at the back. The threshing machine, however, separated the corn into three grades and it was possible to adjust the mechanism to obtain the quality that was required from the grain going through the machine. Further, there was another chute where the weed seed, that had been separated out from the corn, was bagged off and of course the straw, chaff and cavings were all separated. What is more, the straw was not smashed into small bits and was clean, straight and in many cases suitable for thatching.

Straw, in the days of a far more general and mixed husbandry, was an important commodity. Now it is largely looked upon as trash to be burned in the field where it has grown, robbing the land of much-needed humus and causing devastation to whatever insect and other wildlife is able to withstand the ravishes of numerous and varied chemical sprayings. In years gone by straw was a valued asset, not only for thatching and feeding, but

136

also, most importantly, as bedding for livestock, eventually to be returned to the land as farmyard manure. In the days of organic farming, in spite of what the bureaucrats and laboratory farmers might like to say, food must have been less contaminated than it is today by sprays and fertilisers, the long-term effects of which are, in many cases, not truly known in spite of all the claptrap spouted by these worthies.

One use of straw, that I well remember with pleasure, was 'setting the stable fair' at Britwell on a Sunday morning. Edward Clifton-Brown used to like to call in and see the horses on his way back from church. Charlie Hawkins, the head carter, would go out to the stables about halfway through the morning and top up the bedding with fresh wheat straw in each stall, bringing it a good eighteen inches up the sides. He would then lay a 'mat' of straw the full length of the stable, three feet out into the passageway and, with the aid of a fork handle, fashion a straw rope along the front, setting the stables off to perfection. Having done this he would, with a water-brush, work an intricate pattern of diamonds on the quarters of all the massive Suffolk Punches that occupied the stalls, portraying the true beauty of horsepower. No one asked Charlie to do this; he did not look for overtime. He was proud of his charges and wanted 'the boss' to know and see that he cared.

When Jack and Tom had set up the thresher and had everything ready, the former would give it a run to see that the tractor was correctly aligned and all was working to his satisfaction. When he was sure everything was running smoothly, the drum humming with that unmistakeable drone that foretells hours of toil, he would stop the tractor and final arrangements would be made. First, a length of wire netting, held in place with thatching pegs, would be run out around the rick so that it was totally enclosed. Then, and only then, would the thatch be removed, the pegs being gathered up and tied in bundles. It was not unusual for rats to be lurking just under the thatch and to go hurtling over the side to the ground, where they found their escape blocked by the wire. Unless they were quick enough to turn back into the rick they would be despatched immediately by a member of the threshing gang. If dry, the thatch would be used to make a base for the straw rick. When all was ready Jack would fit the starting handle into the big fly-wheel of the tractor and, with the aid of another man, turn it until with a cough and a splutter the engine would burst into life. Jack would put the drive pulley in gear and, when sufficient revs were achieved, Tom would start feeding the machine.

There were two chutes for the best of the grain, so that when one sack was full the flow could be transferred to a second, but for seconds and thirds there was only one chute each, as these came off in much smaller quantities. To anyone not experienced it might have appeared that Jack Keen was somewhat aloof from the rest of the gang as he walked around the thresher, the straw elevator and, in later years, the tier or trusser, oil can in hand. In fact he was constantly alert, listening for the slightest change in the steady throbbing drone which would indicate that all was not well and some adjustment had to be made. In addition, he was responsible for taking off the full sacks of grain, putting on empty ones and weighing the corn. This had to be done at the recognised and accepted weight for a sack: wheat was 20 stones, barley 16 stones and oats 14 stones. In all cases an empty folded sack would be placed on the scales with the weights. It was no small task – handling 2¼ hundredweight sacks of wheat on one's own was certainly a man's job!

A constant watch was kept throughout the day for rats: each sheaf lifted might produce one. There was an air of permanent expectancy, for rats were money. The men working on the rick usually had an old bucket or tin with them and, as nests were uncovered, the young ones, too small to run away, would be thrown into the container. All counted, a rat was a rat whatever its size. There was quite an art in lifting the sheaves correctly. They had been laid to a specific pattern and to keep a steady flow going the procedure had to be reversed. In the early days, on a number of occasions, I found myself partially standing on the sheaf I was trying to lift! This may sound stupid, but straw, particularly wheat, was much longer than it is today and a sheaf could easily be four feet or even longer.

I noticed over the years that one usually uncovered rats in the roof of the rick and, after lowering the height by some five or six feet, did not see many until within a few feet of the bottom. I have yet to meet anyone who has a liking for rats in the wild. I have met one or two people who have kept them as pets, but in their natural environment they bring out the killer instinct in even the mildest of people.

One such person was George Gunn. A little man barely five feet tall, a baker by profession, he had been told by his doctor that he had to give up his work in the heat of the bakery and lead an outdoor life. George hated rats. He confided in me that he was scared of them and that they made him 'come over all cold'. Normally nothing would get him near one, but when in a threshing gang – and he would do almost anything to avoid this – he was a different person. He became a tiger! I was on the rick one day with him and Jack Adaway when a huge doe rat, very full of young, scampered across the rick straight for the thresher. George, with a howl that would have done credit to an Apache brave, his fork held out in front like a pole vaulter, charged. When the rat was within about two feet of the edge he drove at it with all his strength, which it must be said was not a lot, but enough to send George flying up into the air before crash-landing on the far side of the deck and shooting over the side. He had missed the rat! Fortunately he landed in a heap of cavings and returned to the rick a few minutes later, shaken but unhurt.

Nearly all the older men regularly wore 'yorks', leather straps holding their corduroy trousers in tight below the knees, although some just used a piece of old binder twine. At the time of which I write, a large majority of manual workers wore corduroys because they were tough and durable. When warmed in front of a fire they were, to put it politely, offensive to one's olfactory organs. Should the wearer have been doing a few days' dung carting, then the smell was truly something to remember! The chief reason for the 'yorks' was to keep the bottom of the trouser legs up out of the wet and mud, but when threshing they served a double purpose – namely to stop rats and mice running up inside. Sometimes a newcomer to farm work would pooh-pooh the old practice and, at times, even the experienced would fail to take the elementary precaution of tying a piece of twine round below the knee, before going up onto a rick. This on several occasions led to happenings that were hilarious for the spectators, but not always so amusing for the main participant.

There was one such event, towards the end of a rick of barley at Lynch Hill, which caused much merriment, but was most certainly not funny for the man involved. He was a casual worker, one of several taken on to help lift the potato crop. In those days potatoes had to be hand picked into buckets, after being thrown out by a spinner, before being bagged, a most back-breaking and arduous job, and one frequently done by women. This man (I have long since forgotten his name) told Tom Rose that he was an experienced farm worker and, for various reasons, being one short on the threshing gang for the day, Tom had told him to go to Lynch Hill instead of to the potato field when he arrived in the morning. He proved to be a good worker and kept a steady supply of sheaves coming all day to Jack Adaway who was the 'cutter'. We had just got down among the rats and the sport was becoming fast and furious when suddenly the new recruit gave a yell, dropped his fork and started to leap around, grabbing wildly at his leg. His hands went higher and higher, before finally clutching at his crotch; at the same time he gave a positive bellow, before starting to yell that he'd been bitten. It would be indelicate to say where! His discomfort was not helped by Jack breaking out the first aid kit, seizing a bottle of iodine and applying it liberally. Rat bites can be very poisonous.

A rather similar incident took place at Britwell when we were threshing a two-year-old rick of oats. It produced little good grain but an abundance of rats – over 400 of them when one included 200 babies picked out of the nests. The last sheaves were being lifted and there were rats in every direction inside the wire. I had arrived about an hour before, with Bridget and my ratting stick, which had been fashioned by Charlie Davis. It was curved rather like a hockey stick, but flat and about three inches wide at the business end, tapering up to a well-rounded handle. It was a vastly superior 'rat basher' to a straight stick. Bridget was in her element. At one stage I saw her holding down two half-grown rats with her front paws, while she despatched another with a vigorous shake. She never wasted time. Unlike some dogs, who will go on shaking a rat long after it is dead, she knew when it was, would drop it and look round for another.

I was at the heart of the mêlée, doing sterling work with my stick, when Jim Brookling, who was not part of the threshing gang and had just arrived at the farm, jumped into the arena brandishing a stick. It was only a matter of seconds before he let out an oath and grabbed his thigh, shouting 'A b . . . . .'s gone up my trouser leg!' Jack Herbert, old Bill's son, started leaping around with a two-grained fork, brandishing it wildly and making lunges in the general direction of Jim's leg. The latter, realising he was in greater danger from Jack than he was from the rat, hastily beat a retreat back over the wire still grasping his leg. He literally squeezed the rat to death and, as soon as it had dropped out onto the ground, Jim was back in the fray.

It was a phenomenal slaughter and the last really big killing of rats on the estate for, shortly after, a German-made gas, 'Taboo', came on the market. It was said to be harmless to humans, although it certainly made one cough and splutter if inhaled. It was very simple to use. One had a gas-gun – a metal tube about 3 feet 6 inches long leading to a container, with a spring-loaded cap, to take the cartridge once the latter was lit. The gas, brown and pungent, was heavy and would lie over the ground, just inches above it, as one moved the end of the barrel from hole to hole. The manufacturer claimed, rightly, that it was equally effective on rabbits; further, that one whiff of the gas and a rat would eventually die. In large doses death took a matter of minutes.

Many of the hedgerows were honeycombed with rat holes and George Devonshire was told to gas all he could find. He quickly discovered that the rats would bolt from the gas, as they would from ferrets. I spent many hours with him, when 'Taboo' gas was first used on the estate, shooting the rats with my .410 as they made a run for it. The gas was an excellent exterminator, but although I tried hard to obtain it after the war, when managing an estate, I could not find any. Perhaps the fact that the manufacturer was a German Jew – he had, he told us, a small family business – had something to do with this. He used to come over from Germany, to supply his customers direct; he was a pleasant, bespectacled little man with a dry sense of humour and a fanatical hatred for Hitler and rats! We never saw him after the spring of 1938.

Very occasionally ricks, when threshed, used to yield more than corn. One of the farm workers, 'Froggy' Hawkins, equally known as 'Grunter' (for if there was any heavy lifting to be done involving a number of the staff, he was renowned for the maximum of grunting and the minimum of effort) lost his watch. It was a matter of great concern to him, for it was an Ingersoll Crown pocket watch: they cost five shillings, were most reliable and lasted for years. He missed it one day during the harvest, but he'd no idea whether he lost it pitching sheaves up onto the wagons in the field, or on the rick in the evening. Some nine months later he was passing sheaves back on a rick being threshed, lifted a sheaf and there was his watch. He wound it up, it worked perfectly and gave many more years' service.

On another occasion, on a crisp and sunny winter's morning, one of the rick gang forked up a sheaf and, as he did so, he just caught the glint of something bright in the straw. Curiosity prompted him to take a closer look and there, nestling in the sheaf, was a heavy gold wedding ring. None of the farm staff had lost one and, in any case, its weight and quality were such that none of them could have afforded such a ring. Its origin remained a mystery. The police, to whom its discovery was reported, felt it must have been stolen, and that the thief had then panicked and thrown it into the field. If this was so then one or two stalks of wheat must have grown up through it. The ring would have to have remained attached to the straw while going through the binder, being stooked, carted and ricked for it to be found where it was. Impossible as this sounds, it appeared to be the only answer. Another find, nearly in the centre of a rick, was a dead cat. How it got there must also remain a mystery!

There was much to be said for ricking corn, particularly once it was possible to obviate the scourge of rats and mice. It was rather like banking it. The farmer, if he could afford it, could wait and thresh a rick as he needed the money, or when prices increased. There were no grain mountains, no problems of storage. The ricks themselves were works of art and monuments to rural craftsmanship.

12

The Flock

A^S the years passed my involvement in the various departments on the estate became more serious; none more so than with the – eventually – world-famous flock of Hampshire Downs. Sheep for many centuries had been the keystone of British agriculture, particularly on light land. Ever since the introduction of swedes and other root crops the development of folding sheep over these meant that many thousands of acres became a viable proposition; prior to this form of husbandry they were virtually non-productive. Certainly sheep were a must for much of the Burnham estate, for in the main it was light gravelly soil – really hungry land!

The flock was founded in the autumn of 1921, soon after father took over the management of the estate. It rapidly expanded to some 250 breeding ewes, 100–150 ewe lambs, 80–100 wethers and, most importantly, ram lambs being run on to sell for breeding in the late summer and early autumn: some 500–600 sheep in all. As can well be imagined, such a number required a large acreage of fodder crops to maintain them and, indeed, several shepherds. Folding sheep is fairly labour intensive. The head shepherd for many years was Harry Wadman, aided by his son Philip, George Wright, whom I have already mentioned, as carter and two field shepherds. The distinction between these men and the Wadmans is that the latter did not trim and prepare sheep for exhibition at the agricultural shows, nor rams for the sale ring. Theirs was the hard graft, out in the fields in all weather, setting hurdles (that is, constructing pens with sheep hurdles for feeding off the crops). This is where I made my debut in shepherding. When moving hurdles the men used to carry six or seven on a stake over their shoulders and lay them out around the area to be enclosed. I found that three was all I could manage and even that, to begin with, made me puff and blow! Each lot had to be carried from the area that had been eaten off and away beyond that being folded to the next section to be utilised, often well over 150 yards, or more. The town dweller, out for a day in the country, who gazes upon the pastoral scene of sheep, feeding contentedly within their fold, could little guess the amount of sweat and toil that has been required to achieve such a picture of peace and rural tranquillity.

Much shepherding is just monotonous routine work: setting hurdles, carting water, paring feet to control foot rot – an ulcerative disease found chiefly in folded flocks – and moving the sheep from one lot of food to another. When the latter took place it was the custom, with Hampshire Downs, for the shepherds to walk in front, the sheep following and the dogs bringing up the rear. At Burnham, as was the case with many flocks throughout the south of England, the dogs were Old English Sheepdogs, the working forebears of the breed now better known for advertising paint than doing the job for which they were bred. Their origin is rather obscure, but they have certainly been around for many centuries. In 1586 Conrad Heresbatch wrote the following description of an 'ideal English sheep-dog':

> The shepherd's masty, that is for the folde, must neither be so gaunt nor so swift as the greyhound, nor so fatte nor so heavy as the masty of the house; but verie strong, and able to fighte and follow the chase, that he may beat away the

woolfe or other beasts, and to follow the theefe, and to recover the prey.

Fortunately at Burnham they did not have to 'beat away the woolfe', but they were strong compact dogs, far removed from the powdered, pampered creatures that one sees at Crufts. They were gentle, efficient dogs, that did their work quietly and with feeling. I well remember one old bitch of Harry Wadman's called Fly, not exactly an original name for a sheepdog. However, be that as it may, at lambing time she was more nursemaid than herder! Most of the ewes were moved into individual little pens to lamb, but not infrequently when lambing was at its peak a ewe would lamb in one of the yards. When this happened Fly, who was a law unto herself and went where she liked, would go and sit by the newborn lamb, giving little woofs to attract attention, until mother and child were moved to a maternity unit! Frequently a shearling ewe or teg (that is one about two years old), when lambing would often move away from its first born, if having twins. If this happened in one of the yards Fly, assuming she had decided to be on duty, would gently nudge the lamb along with her nose, making sure the new arrival did not become separated from its mother.

Lambing was a fascinating time, but it was equally an anxious one. A constant vigil had to be maintained to ensure that, should there be any complications, someone was there to help with the birth. There was always at least one shepherd in attendance day and night. Hampshires, like other Down breeds, lamb early and normally we had a few lambs by 25th December. One year, when in my late teens, I had a desire to see a lamb born on Christmas Day. So far that year there had been no arrivals. I joined Harry at the Sheep Sheds soon after 11.00 p.m. on Christmas Eve. Immediately after I arrived, we walked together through the ewes lying in the well-strawed yards and sheds. The stars shone brightly in the clear sky and frost glistened and twinkled on the ground in the light of the hurricane lantern.

Harry pointed at one ewe. 'Her'll be first, reckon we'll have a lamb for Christmas.'

We went back to the shepherds' room, where a coal-fired stove glowed, giving much-needed warmth. I made the cocoa, while Harry cooked a couple of kippers. After we had completed our repast, Harry enquired how long I was going to stay. I replied that hopefully until a lamb was born, but I had been told I was not to stay after 2.00 a.m. He said that if that was the case he'd get a bit of shut-eye and that I was in charge. With that he stretched out on the truckle bed and was asleep in minutes. I looked at my watch – five minutes to Christmas Day. Quietly I went outside and leaned against the wall. The bells of St Peter's, Burnham, rang out across the countryside heralding Christmas. I stood there listening, surrounded by sleeping and contented sheep. I thought of the baby Jesus, of the shepherds as they watched their flocks by night – the sheer peace and tranquillity of that moment I have never again experienced. For those few fleeting minutes I felt nearer to God than I have ever done in any church. Little did I know that the bells would only ring Christmas in twice more before the world went mad and mankind was precipitated along a road that for millions would lead to death, moral degradation and despair.

My thoughts were interrupted by a baa of distress. I returned to reality, collected the hurricane lantern and moved into the next yard which housed the ewe that Harry had said would be the first to lamb. There she was, stretched out on the ground, the water bag protruding, straining hard, but making little or no progress. It was obvious that she needed help. I turned to go and rouse

Harry, then hesitated. He had said I was in charge. Time enough to fetch him if I couldn't cope! I had watched dozens of lambs being born, it was time I ceased to be a spectator.

I took off my jacket, rolled up my sleeves, disinfected and soaped my hands thoroughly and broke the membrane, releasing the waters. Gingerly I sought the cause of the trouble. It wasn't difficult to find: one foreleg was back. Having worked this forward it was all plain sailing from then on and within minutes a lovely strong ram lamb lay on the straw being licked and nuzzled by its mother. I found the iodine bottle with a feather in it, which was used as a brush, and dressed the navel as I had seen done on many occasions. Having put the bottle back where I had found it, I picked the lamb up by its forelegs. Then, making baaing noises to keep the ewe's attention as she followed, I removed mother and child to a pen. Seldom have I felt such simple and uncomplicated satisfaction as I did that frosty Christmas morn. The bells had stopped and an almost eerie stillness lay across the countryside, suddenly to be broken by the hooting of an owl in the nearby spinney. Slowly I walked through the ewes. All seemed peaceful enough; there would be no more Christmas lambs, at least not for a few hours anyway.

I collected my jacket and went back into the room where Harry still slept soundly. I doubted very much that he had even known I had left. I boiled the kettle and made a pot of tea before waking him, wishing him a happy Christmas and telling him, with great pride, that the first lamb had arrived. Having finished my tea I set off for home. It was pitch dark as I stepped outside, the stars no longer visible and a chill wind starting to blow. As I rode my bike slowly past Britwell, deep in thought and mulling over the happenings of the last two hours, snowflakes began to fall. It was going to be a white Christmas! Content, I pedalled faster as the snow became more intense, sure that God was in His heaven and all was right with the world. How wrong I was regarding the latter.

That year I spent more time with the flock than ever before – the miracle of birth fascinated me and I rapidly became very adept at dealing with any complications. One thing that intrigued me was how Harry, and indeed Philip, knew every ewe as an individual. This was not something unique to them, for any shepherd worth his salt can distinguish between his charges. All the lambs were 'ear notched', that is, a small piece or pieces were nicked out of their ears. The placing of these notches recorded the sire and breeding. An experienced shepherd can see a family likeness in his charges, while to the inexperienced they all look exactly the same.

Many years later, when running a big estate, we had a flock of some 1,200 Cheviot ewes that spent most of their lives on the mountains. One of the shepherds proved to be a remarkable old man. The ewes were lambed outside in the fields. Unlike at Burnham, no one sat up with them, although the head shepherd would visit the most forward ewes once or twice during the night, according to how much activity there appeared to be. It was late in the season and the hoggets or shearling ewes, always the last, were lambing. More than forty had given birth one particular night, at least a dozen having had twins. In almost every case where this had happened the ewe had wandered off, leaving its first born on its own while she produced the second. I arrived on the scene in time to witness one of the most remarkable exhibitions of natural talent I have ever seen. The aged shepherd, Jerry – an Irishman – was quietly surveying the 'orphaned' lambs. I expressed my concern and enquired what we could do to get them reunited with their dams.

Jerry assured me all would be well and not to worry.

Having got his pipe going to his satisfaction, always a lengthy procedure, he caught a lamb with his crook, put it under his arm and walked slowly off through the ewes. After a few minutes he stopped, put the lamb down by a ewe that already had one offspring suckling her and it was immediately accepted. (It should be pointed out for the benefit of those who don't know that a ewe will only accept and suckle her own lambs and will butt any interloper out of the way.) I had spent my entire lifetime among livestock but I stood and watched spellbound. Only once was Jerry wrong and he quickly rectified his mistake. After half an hour we had no 'orphans' left. I congratulated him on this remarkable achievement and asked how he managed it.

'Be-gob an isn't it 'asy enough. T'ere's always a family likeness so t'ere is. All youse need is th' eye.' He, without doubt, had the eye. I most certainly had not! I could distinguish between some of the Hampshire ewes, but the white faced Cheviots had me completely defeated.

Of course lambing a big flock meant that one had some casualties. Ewes died and lambs were born dead. When the latter happened the dead lamb would be skinned and the 'jacket' fitted over another lamb, either one of weakly twins, or a real orphan whose mother had died. The smell would be right and the ewe would adopt the changeling in place of her own offspring. After a day or two the skin would be removed and the lamb would be reared with all the loving care the ewe would have shown for her own progeny. When a ewe died and there was no foster mother available, then the cade lamb would have to be reared on a bottle. This is simplicity itself, but has one snag – the orphan thinks that whoever gives it its bottle must be mum! Consequently, if given the opportunity, it trots around after that person baaing virtually nonstop.

One is inclined to think of vandalism as being something which is exclusive to present day society and the product of these undisciplined times. This is not so. It has always existed, but not on its present grand scale. The mid-thirties were no exception. One year it was decided not to lamb the ewes at the Sheep Sheds, chiefly so as to be able to rest the adjoining land. To provide alternative accommodation the corn ricks were all threshed at Lynch Hill and a vast L-shaped straw rick was built. Two rectangular yards were then formed, with the giant rick giving ample shelter on the north and east sides. The two remaining sides were a lean-to type erection made with straw packed between sheep hurdles. A snug, warm and highly inflammable fold! Lambing started two days before Christmas and increased in tempo until reaching its peak, by which time the improvised yards were accommodating some 200 ewes and at least 300 lambs. That was when an attempt was made to burn the flock!

It was the first Friday in February. Bob Hedges and George Devonshire had just returned to Lynch Hill, after collecting their wages from the estate office. It was about a quarter to six and the two men were standing outside the former's cottage, from where they could see across to the rickyard. It was a dry, bitterly cold night, with a southerly wind. Suddenly a sheet of flame engulfed the end of the L-shaped rick. Both men clearly saw two youths who were leaping around, only a few yards from the burning straw, waving their arms and shouting obscenities before they raced off into the darkness, apparently heading for Slough Trading Estate. The two gamekeepers raised the alarm and, with other members of the staff, tore across to the fold. Ironically Philip Wadman, the shepherd on duty who lived only some two hundred yards away, had left his charges not fifteen minutes before to go and have his tea as all appeared to be well.

I was returning from the village when I saw the red glow in the sky and, realising that it was a fire and could only be at Lynch Hill, rushed home and into the house, told father and tore off on my bike. As I arrived the last of the sheep were being driven out of the pens. The flames were racing along the L-shaped rick and had spread to the lean-to shelter. Seconds after I got there the straw, shaken out for bedding on the ground, ignited. Fanned by the wind the fire spread at a frightening pace and within five minutes it became a roaring inferno in which nothing could have lived. Bob estimated it was about seven minutes from the time that he and George saw the start of the fire to when the last of the sheep were driven to safety. There was only one casualty, a day-old lamb which had been trampled to death. Had the two men not been where they were, at the time the fire started, the Burnham flock would have been burned alive. Subsequently a large brass petrol lighter was found by the police, in the area where the two youths had been seen; there was also a can that had obviously contained paraffin. Dozens of people were questioned by the police but the culprits were never apprehended and the motive for such a horrific act was never established.

Another horrendous happening occurred early in April, two years later. The pick of that year's crop of lambs, together with their dams, were being folded on a late piece of kale and swedes in the field that ran from the back of Lynch Hill Farm down to Cocksherd Wood. They were totally screened from the road by the wood and could only have been seen, with difficulty, from Crow Piece Lane. Harry, who was looking after this unit himself, claimed with great pride that they were the best ever bred at Burnham. There were only two more days' feeding left and then these elite of the flock were to be moved to the Sheep Sheds, in preparation for the summer shows and ram sales. Imagine Harry's feelings when, on the penultimate morning, he arrived to find that seven of the best lambs had had their throats cut, been crudely beheaded, skinned and the carcases taken away. I have seldom seen any man more distraught than Harry when he arrived at the house, while we were having breakfast, to break the news to father. As in the case of the fire the culprits were never caught, although maximum police effort was maintained for many weeks. The chief constable, Colonel Warren, a close friend of Edward Clifton-Brown's, even solicited help from Scotland Yard, but to no avail.

However, all was not normally gloom and doom regarding the flock – far from it. Usually there were grounds for much rejoicing and pride in its achievements in the show- and sale-ring. However, it was not until 1935 that the pinnacle of success was reached by winning the Supreme Championship at the Royal Show at Newcastle and four of the five classes scheduled for the breed. What made this show particularly memorable was the send-off the exhibits received. For the first time they travelled in a specially constructed sheep van. This was loaded onto a flat railway truck and was, indeed, one of the forerunners of container traffic! When the time came to set off for Taplow station, a pair of matching Suffolk Punches, coats gleaming and manes and tails braided, were between the shafts. The route taken was via Burnham High Street and quite a crowd turned out to see them. Harry rode beside Charlie Hawkins and many wished him well as they passed. When the van arrived at Newcastle, the railway company had arranged for a team of horses to be ready to take it out to the show ground. Around this time Edward Clifton-Brown was made High Sheriff of Bucks. Such was the fame of the flock that he was frequently referred to as the High Shepherd! He confided in father, as he was coming to the end of his term of office, that while it had been a great honour he was relieved that it was coming to an end. Apparently, had there been a murderer convicted within the county and sentenced to death, he, as High Sheriff, would have had to be present when the execution took place.

By law sheep have to be dipped annually, an arduous but necessary task, and the police notified when this was to take place. It was customary for the local 'bobby' to make a casual call while it was happening, thus invoking the spirit of the law. There would be a bit of banter and the arm of the law would be on his way, but as Burnham grew so did the police force. One year an unknown constable arrived at the Sheep Sheds where the dip was situated. He was very young and most aptly described by a farm pupil, who was helping with the dipping, as being pink and white and smelling of bread and butter! He was also most officious.

He first put in an appearance in the morning and then, to everyone's surprise, returned in the afternoon wearing rubber boots. He announced, to no one in particular, that he was going to see that the job was done properly and proceeded to climb into the holding pen from where the sheep were being pushed into the dip. Definitely a mistake! It was mandatory that each sheep was totally submerged. The easiest way to ensure this was to tip them in tail first, so one grabbed a sheep and ran it backwards to the edge of the dip and over it went. The constable took up a position right on the edge, carefully watching to see that all was done according to the letter of the law. After a few minutes he started to become critical, saying that some of the sheep were not being kept in the dip long enough and he queried whether the mix was sufficiently strong. That was his second mistake. Harry Stanley, one of the field shepherds, grabbed a ewe and let it run backwards, catching the constable just behind the knees and

155

precipitating him head first into the bath! Immediately there was a great furore and everyone went to the rescue of the unfortunate man. It must have been a case of 'too many cooks spoil the broth', because three times he was nearly pulled to safety, but somehow hands slipped and back he went again. However, at last he was out and coughing, spluttering and dripping sheep dip, he climbed on his bike and rode away without as much as a thank you to those who had rescued him.

Next day father met Sergeant Garret, who had participated in many a poacher hunt with us, and said how sorry he was to hear that the young constable had fallen into the dip. The sergeant gave father a wink and remarked that these new cadets were so keen, they were always wanting to look into everything. It was the only time I knew of a policeman being dipped, but sheep dogs were a different matter. They were thrown in every year to get rid of unwanted visitors; likewise at shearing time they were clipped out, leaving them cool and clean for the summer.

There were many moments of triumph and pride for those involved with the flock, but I think for father the ultimate came when, for the third successive year, the Burnham flock won the Prince of Wales Gold Cup, for the Supreme Champion pen of sheep over all breeds at the Smithfield Fatstock Show, something that no other flock had ever achieved before. In those days the show was staged at the Agricultural Hall in Islington and the supreme champions from both the cattle and sheep sections were penned in the centre. Edward, Prince of Wales, was a regular visitor to the show to present his challenge cup and other major trophies. It was customary for the royal party to go first to the winner of the cattle section. It appeared that this year was to be no exception but on catching sight of father, standing with Harry Wadman by the winning sheep, the Prince broke away from the official party and advanced, hand outstretched, to greet him.

'Hullo Twist, you seem to make a habit of this. Don't you ever give anyone else a chance?'

'Not if I can help it,' came the reply.

His Royal Highness stood chatting to father until an equerry, or some such person, came over and most politely suggested that the Prince should revert to the agreed schedule. Father had been amazed that the Prince of Wales had remembered his name, although they had met on quite a number of occasions, but this was nothing to the surprise and embarrassment that was to follow.

The Prince said, 'Don't rush me. I've known this man for years. Why, once he actually knocked me over and swore at me!' Laughing, he put his hand on father's shoulder and said, 'I'll see you again in a minute.'

Poor father, he was most disconcerted, for what the Prince had said was true. It had happened at the Royal Show at Cardiff early in the twenties. Father, an all-round judge of livestock, had been judging a number of breeds of poultry. These used to be accommodated in large marquees, the exhibition pens in two tiers, rows being placed back to back. While judging was taking place no one was allowed in except the judges and their stewards, and even the latter kept well out of the way except when wanted. Father was in the middle of judging a huge class that stretched down one row of pens to the end and continued on up the other side. He was nearing his final selection, and was hurrying round to look at an exhibit on the far side, when he literally crashed into someone on the corner coming the opposite way. This person fell to the ground.

Tersely father enquired, 'Who the hell are you and what the devil do you think you're doing?'

The reply was immediate. 'Well actually I'm the Prince of Wales and I'm just trying to have a quiet look round.

157

Give me a hand up, there's a good fellow.'

Father's embarrassment had been acute. The Prince asked his name and whether he might stay and watch father make his final selection. He also asked that a steward should be sent to the far end of the marquee to tell the rest of the party that he, the Prince, would rejoin them in a few moments. The Prince may have had many weaknesses, but he certainly had a good memory!

13

Opportunities

EARLY in the thirties a further innovation in haymaking became standard practice; the use of the haysweep. This was something like a huge wooden fork, 10–12 feet wide, that was pushed along the ground, gathering up the hay from the rows, sweeping it to the elevator so the hay was ricked in the field. This saved all the effort of carting hay, often long distances, to the rickyard and greatly reduced the time taken to ensure the crop was safe. It also meant a quite considerable reduction in the labour required to clear a field. The sweep was the invention of a farmer in Wiltshire (he also pioneered outdoor milking parlours) who was in search of any means to reduce costs in the dreadful depression that struck agriculture in the late twenties and early thirties. Further, he worked out a simple form of traction for his haysweep, namely an old motor car, modified to start on petrol and then run on paraffin – just as the tractors did in those days. The sweep was attached to the front axle of the car by means of a simple and well-thought-out coupling. Thus it could be pushed along at a great rate, making it both the fastest and cheapest method of gathering hay that had so far been devised. A good old car could be bought for a few pounds and the conversion to paraffin cost about twelve pounds. Right from the time that the first sweep arrived at Burnham, I was number one driver. I had been driving tractors and the estate lorry from the age of about eleven or twelve, both in the fields and around the lanes. I was equally at home behind a steering wheel as I was on my bicycle!

With hindsight it was a time of missed opportunities. Cars that were hammered to death sweeping up vast quantities of hay would be worth a small fortune these days – some more than others. Two that we had at Burnham would certainly have come within this category. The first was a Ford, the next model to be produced after the famous T. It was a two-seater saloon and the log-book showed that it had had one owner, namely Viscount Astor. I rather think it had been custom-built for his lordship. When purchased from Sands' Garage in Burnham it was in immaculate condition, real leather upholstery and not so much as a scratch on the paintwork. It had done just over 38,000 miles and cost £10! 'His Lordship', as the car was dubbed by Tom Rose, did two seasons' hay sweeping before being towed away for scrap. Even then the paintwork was virtually without a blemish, but pushing tons and tons of hay around at speed didn't do a lot for the engine!

'His Lordship' would undoubtedly be a collectors' item today, but nothing to what his successor, 'Mrs Simpson', would be. No question about her not being custom-built. She was a six cylinder two-seater Vauxhall, although one could comfortably seat three in front, even four at a squeeze. The dickey was equally massive. She had been built, I was told, especially for H.R.H. the Prince of Wales. The logbook confirmed this. A truly magnificent car, black and red with vast chrome headlamps, the upholstery was red morocco leather, while the dashboard was the most beautiful polished walnut. She cost £15 and had done about 60,000 miles when relegated to the land. I have to admit to christening her – being purchased not long before the abdication, the name was inevitable – but what a car and what power. I could bring some really big loads of hay to the rick in her, which wasn't always popular with the men pitching into the elevator. In fact, when going all out it took three, sometimes four men to keep 'Mrs S' working to capacity.

A very amusing incident happened when we were going flat out in East Burnham Park one evening, trying to have the field finished by dark. At the far end of the field there was quite a steep slope. In the morning, soon after we had started, I swept up a big load and tried to go up this slope too close to the hedge, where the sun had not yet come round and the ground was still damp from dew. The result was a wheel spin. I backed out and left the load to collect later when it had dried out. We were virtually finished, when I suddenly remember the abandoned load. Shouting to the men, who were about to knock off, that there was another one to come, I leapt into 'Mrs Simpson' and raced off down the field. I swung round at the end of the field and charged up the slope at about 25 m.p.h. Wham! The tines shot in under the hay as I gathered up the great pile and sped on over the brow. I had the top down – 'Mrs S' was a drophead coupé – and I thought I heard a shout, but dismissed it as probably someone on the lane which ran parallel to the field. They shouldn't have been there anyway as it was clearly marked 'Private. No Footpath.'

I accelerated towards the rick, stopped and reversed out of the load. As I did so, to my amazement, a very startled young couple rose from the hay! Talk about the Phoenix rising from the ashes! ! Apparently they had been walking down the lane, saw a lovely great pile of hay just the other side of the hedge and decided it provided a great opportunity to have a cuddle. Their unexpected ride lasted less than two minutes, but it was much longer before the lads could stop laughing long enough to get on forking the hay into the elevator. Their laughter was infectious: the two youngsters joined in, then collapsed back into the hay. Things weren't helped by 'Jummy' Young, who was the rick builder. He kept pointing at the two unfortunates and, with tears running down his face, chortling, 'Who'd have thought it, b . . . . y Ted and Wallis in the 'ay.' He then kept going into paroxyms of laughter, which started everyone else off again. Eventually they quietened down and, as the last forkful was pitched into the elevator, 'Ted' and 'Wallis' walked off to the road.

One summer, for a short while, we had a university student who was studying agriculture working on the estate. His father was a friend of E.C.B.'s and he was supposed to be gaining practical experience on the land. He was a nice enough young fellow in many ways, but had what were looked upon by the farmhands as some very strange ideas. At every opportunity he would start on about human rights and the division of wealth. We were carting wheat from Big Field. Several of the four-wheeled wagons had been fitted with draw-bars. They could be pulled by tractors, or by 'Mrs Simpson'. I had brought the tea out to the field gang in the latter, as I returned with an empty wagon. We were sitting in the shade enjoying this, when Bert – his name was Bertram, as he was forever telling all and sundry – started on his hobby horse: human rights. All went quiet. Then little George Gunn stood up, walked over and looked down at the speaker.

'I don't rightly know what you're on about, lad, but I reckon the only right a man has in his life is to work and that's something everyone should have.' Then, with a chuckle, 'And a good woman, or a few bad 'uns. When you've work then the rest is up to you. I don't expect anyone to give I what I haven't earned. If you'm brainy, then good luck if you can make a bob or two, but if not be

162

thankful for having a job. We were put on this earth to work an' that's what you'm should do. Read the Good Book, that'll tell you what's right and what's wrong. It don't say nothin' about lazy idle sods taking other people's money to live off.'

It was quite a speech. I felt like clapping, but knew it would embarrass George. Bert departed the following day. He left a note at the estate office for father, saying that he had come to the conclusion that there was little opportunity to improve the lot of the farm labourer, the latter word being heavily underscored, and that he felt agriculture held no future for him. He was wrong. When war broke out he declared himself to be a Conscientious Objector – not, I am sure, on religious grounds – and spent the war doing farm work. How do I know? After I was invalided out of the army he arrived on the estate I was running. He was in a gang, sent by the County War Agricultural Executive Committee, to do contract hoeing.

I enjoyed the tea breaks when we were haymaking and harvesting. I would sit and listen to the old hands reminiscing. They had a wealth of stories and knowledge on country matters. I remember old Ted Wilmott, who normally worked in the orchards, was a great narrator and could hold his audience spellbound. Originally, I believe, he came from Colnbrook, about fifteen miles from the centre of London. When he was a lad he frequently accompanied his father to Covent Garden with loads of vegetables. The latter had a two-horse dray and contracted to take produce up to the market. He had four light draught horses and used two on alternate days. This enabled him to trot much of the way. Six mornings a week he would be up at 3.00 a.m. and would normally reach the market around 5.30 a.m. Having unloaded he would head straight for home – unless he knew there was to be a hanging at Tyburn. If this was the case he would take a longer route so as to be present. Ted said his old man loved a hanging, considering it to be rare 'sport'. He took Ted once, when very small, to show him what happened if you broke the law. Ted couldn't remember what the two unfortunates he saw strung up had done – he rather thought it was sheep stealing – but the experience had a profound effect on him. He always claimed that public hangings gave the authorities a great opportunity to show that crime did not pay. Once Ted's father got outside London he would hook up the reins, curl up on the floor of the dray and let the horses take him home. That way he was refreshed and ready for a day's work when he reached home.

Another favourite topic of conversation was 'The Fancy'. Many of the estate staff kept and bred various varieties of rabbits, poultry and pigeons, showing them with great enthusiasm, as they did the produce from their gardens and allotments. Virtually every district had a flower show and a section catering for 'The Fancy'. There was much rivalry at these shows, with long postmortems held on the judging when the cards went up and results were known. Burnham Flower Show, held in Priory Meadow, was around the end of July. As such shows went it was a large one. A big schedule was arranged for the flower and vegetable section, which included jams, home-made bread, cakes and floral arrangements. A correspondingly large number of classes catered for 'The Fancy', including classes for children's pets. It was in the latter that I first exhibited. I suppose I was about six at the time. I had been given a pair of black Pekin bantams a few months before the show, providing me with the opportunity to become involved in 'The Fancy' and starting me out along the road to showing many varieties of livestock. However, they also caused me much trouble in the early days of ownership.

I named my two pets Mapekin and Papekin. The first spot of bother came when I decided that Mapekin was not pulling her weight – she wasn't laying any eggs! Mother, always an early riser, was in the habit of having what she described as 'forty winks' after lunch. She would retire to her room and relax for a while on the chaise longue with a good book, or even have a siesta on the bed. She was doing just that, one hot summer's afternoon, when she was rudely awakened by much squawking from Mapekin as yours truly pursued her round and round her pen with a large stick, shouting, 'I'll make you lay, you b . . . . r.' Alas I had been seeing too much of old Bill Herbert. The outcome was inevitable!

The second time in a matter of weeks that my pets were to bring consternation into my life, entirely of my own making, was when, for the first time, I showed Papekin. This was at the Burnham Show in the children's class for bantams. When father penned him for me early in the morning, I thought there was nothing that could compare with my magnificent exhibit. I could hardly keep still, and waited anxiously while the show was officially opened at 2.00 p.m. by Colonel Hanbury, Mrs Clifton-Brown's brother – I thought he'd never stop talking, but at last he did and when he pronounced the show open, I was gone. I didn't wait for the clapping to stop – in fact I hardly waited for it to start! I raced over to the marquee and round to Papekin's pen. There, to my horror, was a blue card – second prize! I was livid and I let my feelings be known. Father gave me a frightful ticking off for being a bad loser and, apparently, abusive about the judge, but I have forgotten that bit.

The unfortunate girl who was also there to enjoy all the fun of the fair (and there was quite a big one) was located and told to take me home. I don't think she was very pleased, but was considerably mollified when told that, under the circumstances, she could return to the fair after supper, providing she was back at the house by 10.00 p.m. She definitely perked up, for she was very attracted to the opposite sex. Old Bill Herbert used to say that there were more bones made in Priory Meadow on Flower Show night than were ever broken! Be that as it may, home I went and, once more, to bed. I was mortified. Not only had my beautiful bantam been beaten, but I was missing all the fun – the swing-boats, dodgems, the roundabouts, all the things I'd been looking forward to for weeks. It was a lesson I was never to forget. I can truthfully say that since that day I have been a good loser and always accepted the judge's decision without a word – even when I have known it to be totally wrong!

Several of the older men on the farms, who were coming up to or had reached seventy – there was no automatic retirement at sixty-five – had been drovers in their youth. This entailed walking hundreds of miles driving sheep and cattle from one part of the country to another. There were quite a number of prosperous farmers who had started life this way, having avoided the temptations of drink and wenching while on the road and saved their money. However, those who had not been so circumspect certainly seem to have enjoyed themselves. Old Bill Herbert, whose tales never lost anything in the telling, had once gone on the road and, given the slightest encouragement, would happily recall his experiences.

It was in the early 1870s, not long before he joined the army, that Bill and two friends had agreed to drive some fat cattle from near Aylesbury to a farm at Hounslow. There they would be rested before their final journey to Smithfield Market. He was told not to hurry the beasts and walk all the fat off them. This suited the drovers fine, for they were being paid 3s 6d a day – a very good wage! The second day on the road, having ensured their charges were secure for the night, Bill and his companions made their way to a nearby hostelry for food, drink and whatever other entertainment they could find, before searching for a place to kip for the night.

Bill felt he was in luck when he picked up a girl in the tavern. He quickly discovered that she was a scullery maid at 'the big house' and, contrary to all the household rules, was in the habit of slipping out of an evening to enjoy herself once her work was finished. Bill plied her with drink and quickly learned that she had a minute room to herself and a bed that she was more than willing to share with him. This was an opportunity not to be missed – a far more attractive proposition than dossing down under a hedge for the night or, with luck, finding some farm building in which to sleep! Bill told his companions that he would meet them in the morning just before dawn and, with that, he was on his way.

The bed was so comfortable and the company so enjoyable that he overslept. Both were terrified when they awoke to find it was daylight. Hastily, the girl took him through the maze of corridors leading out to the back of the house; she was blubbering as she went and told Bill, between her sobs, that if they were caught the master would nearly flay them alive in front of the rest of the staff. He was, she said, a rabid puritan in his outlook and a great one for making an example of those who had sinned! At last they were at the door and the girl pushed Bill through. He looked around and realised he was in a stableyard. The next thing he knew was that two grooms were racing down the yard towards him. He fled. He ran as he'd never run before and, eventually, eluded his pursuers. He was still gasping when he rejoined his mates, who were just starting to get the cattle onto the road.

Several hours later, beginning to feel relaxed and starting to brag about his exploits, he heard hoof beats. A vehicle, being drawn at speed, was coming up behind them. Bill and his helpers herded the cattle off the road, to let whoever was in such a hurry pass. But they didn't. Instead a wagonette stopped beside them and three Peelers leapt out and grabbed the drovers. 'That's him!' Bill realised there was another man present – it was one of the grooms who had chased him. Poor Bill was unceremoniously dragged off, charged with robbery and thrown into a cell. Apparently, while Bill had been in the arms of the scullery maid, the big house had been broken into and many valuable pieces of silver stolen. Bill was terrified, for theft could be a hanging offence, a fact the Peelers were not slow to point out. They said that if he admitted his guilt and told them where he had hidden the loot, he might just get off with a jail sentence.

Two days he remained shut up, then, suddenly, he was released. Apparently the real robber had been caught with the stolen silver. As he was leaving the police station, half expecting some form of apology, the sergeant gave Bill a vicious blow to the side of the head, knocking him to the ground, and told him not to trouble them again! That was Bill's one and only go at being a drover. He never caught up with his mates and never got paid. He claimed that fighting the Zulus was far less frightening!

One drover who did make good was Ben. At least that was what we knew him as, but he seemed to answer to a variety of other names as long as there was money in it. Normally he worked the sheep fairs in Hampshire, Wiltshire and Dorset, but where there was work he'd be there. He was a great character, shabbily dressed, and he always wore a green bowler hat – green from age! I remember being at one fair where father bought some ewes that had to be railed to Taplow. Ben was located and a price of threepence a ewe was agreed to take them to the station and load them. Having paid Ben and tipped him, we drove off across country to see a friend of father's. As we were returning, several hours later along very secondary roads, we came round a corner and there was Ben. A very transformed Ben, being helped into a smart coat by a uniformed chauffeur, standing by an equally smart and opulent car! Father stopped and got out, telling me to remain where I was. I could see Ben chuckling as the two of them talked. The chauffeur had retired respectfully to a distance. It must have been a good twenty minutes or more before father returned to the car. He told me that Ben had given him a summary of his life.

Apparently his father had been a farm labourer. When Ben left school he was unable to find a job, so he started going to fairs and markets, picking up work droving. He saved every ha'penny he could and quickly earned a reputation for being reliable. Soon after the 1914–18 war he attended a farm sale, the tenant having become bankrupt. It was a terrible day, snowing hard and very few people attended. Ben carried his worldly wealth in a money belt around his middle. This was to be his opportunity. There were three large ricks of hay. There wasn't a bid for them, but the auctioneer said they had to be sold. Ben offered £10, but the auctioneer would not accept Ben's bid until he saw the colour of his money. He produced the required sum and they were knocked down to him. That was his first major break, for after a very hard winter hay was in desperately short supply. Ben made what amounted to a small fortune. From then on he never looked back, but he always remained loyal to the work that had given him his start in life – droving. At the time father talked to him on the road Ben owned several farms and lived in a large house with servants. He was not married. He bought and sold hay throughout the southwest of the country and was, in fact, a very rich man.

A few weeks after our encounter with Ben on the

by-road we were at Weyhill Fair, foremost for the sale and letting of Hampshire Down ram lambs. Ben, as usual, was there. He came up to father, touched his old green bowler respectfully and asked if there would be anything for the station. Father said no. Ben stood there. With a wry smile father put his hand in his pocket and tipped Ben two bob. The latter spat on it, put the coin in his pocket, touched the rim of his old bowler again and with a 'Thank'ee sir' put it in his pocket and walked away in search of other clients. By such methods are fortunes made!

14

The End of an Era

IT was early in May 1939 and the sun was shining brightly as I walked along the road from Britwell to Cocksherd Meadow, passing the orchards rich with blossom and the buzz of bees, the latter as yet safe from sprays and man's destruction. The rhythmic clip-clop of the horses' hooves on the hard road gave a sense of peace and security. It was a Saturday. The cart horses were to be turned out to grass for the weekend, the first time that spring. There were four of us escorting the eight massive Suffolk Punches. It did not require that number, except possibly to release them for their first mad gallop after months of restriction; we just wanted to watch when they gained their freedom. I led Boxer and Sprey, rich dark chestnuts in magnificent condition, their coats glistening in the hot sunlight. Both were well over three-quarters of a ton of well-muscled power and yet both were as gentle as kittens.

The estate lorry passed, Harry Jaycock, the driver, giving us a cheery wave. Boxer snorted and pulled away, eyes rolling, showing fear as well he might. It was barely eighteen months since he had been hit by a lorry. It had come down the hill from the direction of Leas Farm towards Cocksherd Wood, and the brakes had failed. Jim Brookling, returning from Big Field with Boxer between the shafts, reached the junction of the two roads at the same time as the lorry. The latter crashed straight into the side of the cart, knocking both horse and vehicle over and catapulting Jim into the road. Miraculously no one was killed, but Boxer sustained a three-cornered cut high up on his hindquarters.

I came along just after he had been freed from the wreckage. A large lump of flesh, which must have weighed at least 3–4 pounds, hung down from a gaping hole. Unbelievably there did not appear to be another mark on him, not even a scratch. Asking me to hold Boxer, Jim walked round and literally folded the lump of flesh back from whence it had come. It fitted exactly, rather like putting a cork back in a bottle. There was very little blood. Quietly I led the terrified horse back to the farm, while Jim walked beside him with his hand on the wound. Luck was with us on two counts. First, we did not meet a motor vehicle of any class and, second, when we reached Britwell we found that Aubrey Ward, the vet, was there. He'd been called out to a cow with milk fever. He quickly examined the gelding and then deftly stitched the wound, doing such a wonderful job that six months later there was virtually no scar to be seen. Alas, however, there was one on Boxer's mind – he was terrified of lorries and so was no longer used on the roads.

We arrived at the field. Having entered and closed the gate behind us we spread out, about twenty yards apart. The horses stood tense and expectant, for they knew as well as us what was about to happen. At a given signal from Charlie Hawkins we slipped off the halters. Boxer reared, his massive hooves only inches above my head, before he wheeled and, bucking and kicking, thundered off with his companions. A wonderful sight! Those massive creatures, some six tons in all, were squealing with joy, jumping all four feet off the ground and capering like spring lambs. The ground shook as they raced away, to do two circuits of the field before finally slowing to a trot at the far end. Wheeling, nearly in line, they came up the centre, tails held high, heads tossing, nostrils dilated as they snorted in the sheer ecstasy of their freedom. Then, almost as though on command, down they went and rolled and rolled, their massive power-filled legs waving like demented shadowboxers. The ring of metal filled the air, as iron-clad hoof struck iron-clad hoof, and all the frustration of a winter's stalling was released. Sprey was the first to rise and, after a mighty shake, she started to graze. One by one the others followed, until all eight had their heads down enjoying God's bounty.

Charlie and the other two men departed, but I sat on the gate watching and deep in thought. War, I felt, was imminent. I did not believe the recent 'exclusive', in one of the more sensational newspapers, that claimed the multitude of German tanks were but cars with a plywood superstructure! So sure was I that this was nonsense that, together with a number of friends, I'd joined the Territorials some months before. Many pooh-poohed the idea of war and said that mankind could never again be so mad. I was not convinced. Their track record did not exactly inspire confidence! Neville Chamberlain, in my book, was a stupid, gullible fool; the cabinet, I thought, were equally ineffectual – like a lot of ostriches burying their heads in the sand. Only one man of prominence appeared to realise where Baldwin's and Chamberlain's policies of disarmament and appeasement were leading us – Winston Churchill – and he was dubbed a warmonger! Vaguely I wondered if the eight gentle creatures, grazing so serenely, would be commandeered by the army.

Alas, however, war or no war, the days of the heavy horse were numbered. They were rapidly being replaced by tractors, as indeed was dung by chemical fertilisers. The unbalancing of nature was truly launched, the decimation of the countryside inevitable and all in the name of progress. The question was – was it progress? I had for years heard father say that the chemist and the inventor were the greatest pest with which countrymen would ever have to deal. That, being townees, they would happily sacrifice the environment and, indeed, our heritage, all in the name of progress, which, quite simply, should be spelt *greed* – man's greatest failing!

I climbed down off the gate and wandered over to the contented munching group. None of them tried to walk away. I gave each one a pat and a lump of sugar – two for Sprey and Boxer, they were my favourites. I leaned against the former, my arms resting on her powerful withers as, after a careful examination of my pockets to make sure there was no more sugar, she returned to cropping the grass. Life seemed so good, with so much to come. I started my final exams at the university in a few weeks' time, but for what? To be blown to bits by some Nazi shell? Or would I go to South Africa at the end of September, as planned, in charge of a consignment of pedigree stock being shipped out there by Harry Hobson & Co.? Sprey took a step forward and literally jolted me back to the present.

Two larks were trilling overhead and a pigeon cooed from the shelter of a nearby fir. I glanced at my watch – it was twenty minutes to one. Lunch would be nearly over and I would not be popular with mother. But still I lingered, loth to leave. It was as though some unseen power held me, etching the tranquillity of the moment upon my mind for ever. A common enough sight, heavy horses grazing in a field, one viewed by thousands over many centuries – but for how much longer? At last I forced myself away. I felt depressed. It had been like a premonition of things to come that I did not want to accept, but knew I must. As I reached the road I did not look back. The picture would remain with me always:

peace, a countryside free of pylons and as yet unsullied to any great degree by the hand of man.

That afternoon I could not settle to revision as I had intended. So, putting my books to one side, I took a gun and accompanied by Sally, father's bull terrier, set off down the lane. It hadn't changed since first I'd been along it in the early twenties, except the trees were bigger. Wildflowers abounded in the hedgerow and a host of butterflies and bees winged their way from flower to flower. I crossed the main road and continued along Lynch Hill Lane. Halfway down the hill I climbed up the bank by the big oak, the one the owl flew out of the night the red dog attacked me, and so along the edge of Lamas Wood. I went past the place where we caught Jim Sloane long-netting and the firs where I shot the five pigeons. I could have bagged several rabbits as they scuttled into the wood; that is what I'd set out to do – shoot two for the ferrets. I let them go. There would be no trouble getting a couple on the way home.

I reached the end of the wood and headed for Hay Mill Pond – or what was now left of it! The pond and reed beds once covered nearly twelve acres. First came the lower pond, wide, deep and providing the head of water to drive the mill. A narrow passage led through the reeds to the top pond, which was only a few feet deep. On a fine day one could drift slowly in the punt and see huge tench, great shoals of roach, and pike, some of which must surely have been twenty pounders. Beyond this a twisting narrow channel brought one to the springs. There were five of these, deep and clear. They were probably much as they had been in 1265 when Burnham Abbey was founded by the Earl of Cornwall, who endowed it with lands and a water mill, south-east of the village, known as Aymill. The stream that powered this flowed on through the village of Cippenham, providing a constant water supply, before finally joining the Thames. For at least five hundred years the springs had been the source from which the water came to operate the mill. Over the years, slowly, by erosion, the mill pond had been formed. In the seventeenth century it is known to have been large enough to provide a goodly supply of fish for its owner, who also used it for keeping swans and breeding cygnets for his table.

When Slough Trading Estate was developed, water was needed for the factories. In the thirties a reservoir was built and a bore, some 1,100–1,200 feet, was sunk to fill it, tapping the underground rivers that fed the springs. Suddenly the flow of water to these was greatly diminished, at times being reduced to a trickle and finally, for short periods, drying up altogether. For the first time ever it became possible, in high summer, to walk on the baked mud between the springs. I was the first to discover this.

One very hot summer's day I made my way through to the boat landing, situated by the entrance to the springs. To my amazement and horror the channel was dry. I retraced my steps, walked out along the duckboards, now no longer needed, leading to the high reeds that screened the springs. I parted these. It was unbelievable, the water had receded so that there were five deep clear pools. The mud between them, usually covered with some two feet of water and estimated to be about twenty feet deep, was dry and hard, at least on the surface! I eyed the baked, cracked silt, then gingerly I tested it: clutching some reeds, I stepped onto it. It took my weight without so much as an indentation. Still clinging to the reeds, I jumped up in the air and landed with as much force as I could muster – there was no sign of the crust breaking. Cautiously I ventured further. All was well.

I relaxed and started to look around me. I had visited

the springs scores of times in the *Dabchick*, a broad ten foot punt built by Charlie Davis, but this was seeing them from a different angle. Still with a certain amount of apprehension I made my way out between the two nearest springs. I looked down into the deep clear holes and wondered at all the thick black sticks protruding from the sides. As I watched one appeared from the mud until, like the others, it jutted out for about eighteen inches. Suddenly I realised that they weren't sticks, but eels! I looked in the other pools, which were all the same. There were hundreds and hundreds of eels, all big. I retraced my steps and then hurried home.

To begin with it was obvious that father did not believe me when I told him I had been walking around between the springs. Further, he was very sceptical about the number of eels I claimed to have seen. When I eventually convinced him, he berated me for my stupidity. Did I not realise that had the baked top layer given way I would never have been seen again? At last he calmed down and agreed to come and see for himself. Muttering something about riparian rights, he got into the car and drove us to Hay Mill Meadow and across the field to the gate leading into what was known as the Pond Tail, which was also the shortest route into the springs. Having got over his surprise, father realised that the eels could be a source of revenue – the estate had to pay its way.

We went back to the office and he telephoned Macfisheries in Maidenhead. The manager was most interested and said he would contact head office. He called back within ten minutes to say they would take all the eels available. Father made up four gaffs, from hooks that we used when fishing for conger eels when we were on holiday. He filed off the barbs and then bound the shanks to stout bamboo canes about twelve feet long. The next day father and I, together with Jummy and Bob Hedges, headed for the springs. Harry Jaycock brought the lorry down across the field and a good supply of strong sacks. We set to work to gather this unexpected harvest, gaffing the fish as near to the head as possible. It soon became apparent that it was essential to have someone to hold the sack up, otherwise those eels already captured wriggled out while we were trying to put the next one in. So we worked in pairs, Bob and I, and father and Jummy. It was quite hard work and we took it in turns doing the actual gaffing. As a sack was securely tied Harry carried it out to the lorry. Some of the eels took fright and disappeared into the mud, but when we'd finished we had filled fifteen sacks. Macfisheries paid for just over 7½ hundredweight of eels. I had taken a spring balance with me and the largest was 4 pounds 5 ounces. They must have averaged around 3 pounds.

The next day I returned to the springs and the eels were still there, protruding from the mud, nearly as many as there had been when first I saw them. I counted several hundred. Father said he hadn't time to do anything about them until the following day. That night there was torrential rain and a frightful thunderstorm. When we arrived at the springs the next morning, it was dull and overcast and the mud was no longer safe to walk on, so that put an end to our plans. However, we could see into two of the pools – and there wasn't an eel in sight!

That had been two or three years before I made my way towards what remained of the pond that May afternoon in 1939. Over the intervening years there had been much correspondence and legal exchanges with the Slough Trading Estate regarding riparian rights, but it seemed that industrial development took precedence over centuries of history, heritage and, indeed, the environment. Such things, in the eyes of the developers, were of little consequence. For, as father insisted, they were not countrymen,

knew little of God's true gifts and looked upon the country
as somewhere to spend leisure hours and leave their picnic
rubbish!

I reached the Pond Tail and started to walk along the old
hedge that ran some twenty yards out in the field, parallel
to the reed beds. Two three-quarter grown rabbits made a
run for the warren under the big holly tree. They never
made it. Sally retrieved them to hand and I hocked them
and hung them up to collect later. Roasted and basted with
milk they would be delicious. I walked out to the landing. A pathetic trickle meandered through the buttercups,
rushes and irises that now flourished on the rich silt where,
until recently, great pike had swum and water rails had
scuttered across the surface, looking as though they really
could walk on water.

I thought of the pike that lived in the lower pond,
between the two mooring poles. I caught him first when I
was ten years old – he'd weighed 2½ pounds. Father had
marked the dorsal fin by making a minute hole through it
with a punch, normally used for marking day-old chicks
between their toes. The pike grew and grew. Each time he
was caught he was marked, weighed and then returned to
his habitat. Once he broke my cast and I lost my favourite
spinner. The following Sunday Ralph caught our friend.
There was my spinner, still attached to his lower jaw.
Gently he removed it before he returned 'Poles' to the
water. (We called him that because he seemed to live
permanently between the two moorings.) I had last caught
him the previous February, with the very spoon that Ralph
recovered. 'Poles' had become a great brute of 16½ pounds
with little space left to mark him he'd been caught so often!
I've no idea to what age fish live, but he must have been at
least twelve years old by then. The biggest pike we knew of
that had been taken from Hay Mill Pond was 21½ pounds.
It had been mounted and graced the wall of the fishing hut

for many years, before being removed to Pound Cottage when the hut was dismantled.

I returned to the meadow and made my way along the hedge. Past the elm tree where the carrion crow had nested and on to where Major Cracroft-Amcotts, Edward Clifton-Brown's son-in-law, had shot the mallard. It was years before, soon after he had married Rhona. He was told to go along the hedge to a certain point and wait until a beater appeared. As soon as this happened he was to walk along slowly, keeping about thirty yards ahead of the beater, until he was in sight of the next gun. The poor major, whom I'm sure only shot to please his father-in-law, seemed unable to grasp what was required of him. I volunteered to show him.

We took up our position and within minutes two mallard came flying up the line of the hedge, directly behind each other. A space of some 8–10 feet separated them, and they could not have been more than, at the most, 60 feet off the ground. They would have been well within the range of my .410. Just before they reached us I realised that the major had not seen them. I cried out 'Over, over' – no doubt in a high falsetto voice, so great was my concern. Up went the major's gun and down came a duck – the second one. 'Good shot.' I ran out and picked up the duck, which was stone dead.

When I returned to the hedge I enquired, 'Why didn't you shoot the first one, sir?'

The major smiled. 'I'll let you into a secret, Michael. I was aiming at the first one and killed the second, but don't tell anyone.'

Poor old Amcotts, he was without doubt the worst shot I ever saw. Once, on the first big pheasant shoot of the season, the major drew 'the hot seat' at Cocksherd Wood. In fairness to him, he begged E.C.B. to let him change places with the left hand gun, for the latter was unlikely to have many shots. However, this was not to be. Father, who was not shooting, stood with the major. Pheasants streamed over and, after at least a hundred cartridges had been expended, there was still nothing on the ground. The drive was nearing the end. Father picked up a stone and aimed at a low flying cock – and down it came! The two neighbouring guns, who had seen the incident, gave a cheer. Bridget, who must have been feeling as frustrated as father, retrieved it. A few minutes later the drive was over. The major carefully blew the smoke out of the barrels of his gun and, with a big smile, picked up the pheasant and walked off with it, remarking to father as he went, 'Well, we've got one Twist. Lovely show of birds.'

I reached the gate and barbed wire entanglement, where Jim Sloane's companion had got hung up. I was about to retrace my steps when, as I looked out across what was left of the lower pond, a pen and a very late brood of cygnets glided into view. Swans had bred on the mill pond for centuries, but soon there was to be no water for them, for within two decades the pond would be dry and built on. The mill was destined to become a roadhouse and, where the cygnets swam, a modern wonder would replace them – a bowling alley! So much sacrificed for so little. I made my way back along the bottom side of Lamas Wood, cut through by the pond where the wild duck used to be reared and came to THE stile. I took both cartridges out of my gun before stepping over! Between there and the lane I shot two adult rabbits for the ferrets, much to Sally's delight, and wended my way home to my books.

Exams were over, results known and the harvest was safely ricked. We, that is father, mother and I, headed off to Cornwall, although the rumblings of war grew ever louder. Ralph had departed to India two years before, as an assistant manager on a tea garden. He hadn't really any interest in tea when he went. It was his love of wildlife and the works of Rudyard Kipling that lured him from his native shores.

Both father and I were avid sea fishermen. It had always been my desire to have a day's fishing off the Wolfe Lighthouse, halfway between Land's End and the Scilly Isles. We'd arranged to go on several occasions with Joe Madron and Ben Jefferies, with whom we regularly fished out of Mousehole, but the weather had never been good enough. It had to be very fine to go all that way in a thirty-five foot open boat!

However, this year, at the beginning of the last week in August, we were lucky. The forecast was good, the sea flat calm. We reached our objective around mid-morning and anchored close to the rock. Within minutes of getting my line out I caught a 16 pound ling. Sport was fast and furious. The lighthouse keepers came out onto the concrete landing to watch and had a shouted conversation with Ben and Joe. I learned from this that the three men were due to be relieved that day, in about an hour's time, having spent two months on the rock. Around 2.00 p.m. the Trinity House boat appeared. It was so calm that they could actually go alongside the landing, a very rare happening! However, there was no sign of any replacement crew, although a large quantity of stores was unloaded. Then the relief boat came across to where we were anchored. Their news was grim: they said war was imminent and strongly advised us to up anchor and head for Mousehole at once. Ben, or Bloody Ben as he was always known, because he prefaced everything he said with 'bloody', was all for fishing the tide out. We'd have to move anyway in a little over an hour, when it changed. Ben received unanimous support.

It had been a wonderful day and, as we steamed towards Land's End, I wondered when next I'd fish those waters. It was a gorgeous sunny evening and we were all relaxed having a smoke when, suddenly, some three hundred yards off our port bow, the sea seemed to erupt. The next thing two submarines appeared out of the depths. It was, to say the least, a bit scary. Ben, an elf-like person, started leaping around waving a gaff and shouting at Joe to ram the b . . . . y b . . . . . s. He was sure they were German! They were not. As they drew level the hatches opened and several seamen came on deck and waved cheerily. The following day we left for Burnham.

The ensuing week was one of disbelief, tension and anxiety. Chamberlain waffled on about Herr Hitler being a man of integrity. Only the ostriches believed him! Girls who had always said 'No', it was claimed, were now saying 'Yes', for next week we might all be dead. But life had to go on, livestock cared for, land tilled. On the afternoon of Friday 1st September I took a load of wheat from Lynch Hill to Leas Farm on a trailer pulled by 'Mrs Simpson'. After unloading I walked across the lane, in front of the house, to the old orchard. There was a very ancient plum tree in it, no one knew how old, nor the variety. The fruit was golden and absolutely delicious, but there were never more than 12–20 plums on it, always at the very top – not worth considering commercially. For some years past I had made the somewhat hazardous climb and picked them. I ate two or three sitting in the top of the tree, before gathering the remainder and taking them home. It had almost become a ritual. This year was no different. As I sat there I could hear the radio in the house. Suddenly the dance music was interrupted, for what was described as an urgent announcement. I listened carefully. All Territorial units were to report to their drill halls immediately. Slowly I ate the last plum, savouring every morsel.

So it had come at last – so much for the meaningless assurances of Chamberlain and his cabinet. As I drove slowly back to Lynch Hill to put 'Mrs Simpson' to bed – it was a standing joke, we never garaged her, we always put her to bed – I thought of father's friend who'd told me that one of the things I would never see or meet in life was an honest politician. By that he meant one who told the truth. I wondered if Chamberlain came in that category, or was he just plain stupid? I opted for the latter. I did not go straight home to change into my uniform, I went to Britwell. Sprey was in the stable. She whinnied softly as I entered, her velvet muzzle brushing my cheek before inspecting my pockets for sugar. I leant against her manger, rubbing her gently between her ears. How generous, loving, hard working were horses – oh, that man could be their equal.

Next day the Battery took over Taplow Grammar School. On Sunday 3rd September, after church parade, the C.O. informed us that war had been declared on Germany at 11.00 a.m. that morning. Further that Lady Astor had come to address us and take the salute as we marched back to 'barracks'. In strident tones she informed us that she had come to cheer us on our way. That, as in the last war, Cliveden, the Astors' stately home overlooking the Thames, would be turned into a hospital for the wounded. She concluded her address with: 'By doing this I am making my contribution to the war effort. You must make yours and I greatly look forward to meeting many of you again at Cliveden in the not too distant future. God save the King!'